STUDENT UNIT GUIDE

A2 Geography UNIT 4

Edexcel

Specification B

Unit 4: Global Challenge (Population and the Economy)

David Burtenshaw

Philip Allan Updates
Market Place
Deddington
Oxfordshire
OX15 0SE

tel: 01869 338652
fax: 01869 337590
e-mail: sales@philipallan.co.uk
www.philipallan.co.uk

ISBN 0 86003 696 0

This Guide has been written specifically to support students preparing for the Edexcel Specification B A2 Geography Unit 4 examination. The content has been neither approved nor endorsed by Edexcel and remains the sole responsibility of the author.

Printed by Information Press, Eynsham, Oxford

P00084

Contents

Introduction

■ ■ ■

Content Guidance

■ ■ ■

Questions and Answers

Introduction

About this guide

The purpose of this guide is to help you understand what is required to do well in **Unit 4: Global Challenge (Population and the Economy)**. The guide is divided into three sections.

This **Introduction** explains the structure of the guide and the importance of finding linkages between units. It also provides some general advice on how to approach the unit test.

The **Content Guidance** section sets out the *bare essentials* of the specification for this half of the unit. A series of diagrams is used to help your understanding; many are simple to draw and could be used in the exam.

The **Question and Answer** section includes three sample exam questions in the style of the unit test. Sample answers of differing standards are provided, as well as examiner's comments on how to tackle each question and on where marks are gained or lost in the sample answers.

Linkages

When the specification was developed it was envisaged that students would make linkages between Units 4 and 5. There are three possible ways in which your A2 course has been organised:

- You are studying Global Challenge as your first A2 unit, for examination in January.
- You have studied Global Futures (Unit 5) before starting this unit, in which case the links to Unit 5 shown opposite are most important.
- You are not allowed to enter any examinations in January and are studying Units 4 and 5 in parallel, in which case select your Unit 5 options to provide the greatest help to your Unit 4 studies.

In the third case, you will be able to make use of the linkages shown opposite.

As you study this unit, you should make sure that you make linkages wherever possible between Natural Environment and Population and Economy. Section C of the unit test contains questions that overarch the whole of Global Challenge.

There are links to your AS course. In Unit 1, for example, you studied small-scale ecosystems in river and coastal environments; this links to global biomes and biodiversity. Similarly, your study of the growth of millionaire cities for Unit 2 links with both international migration and globalisation of the economy.

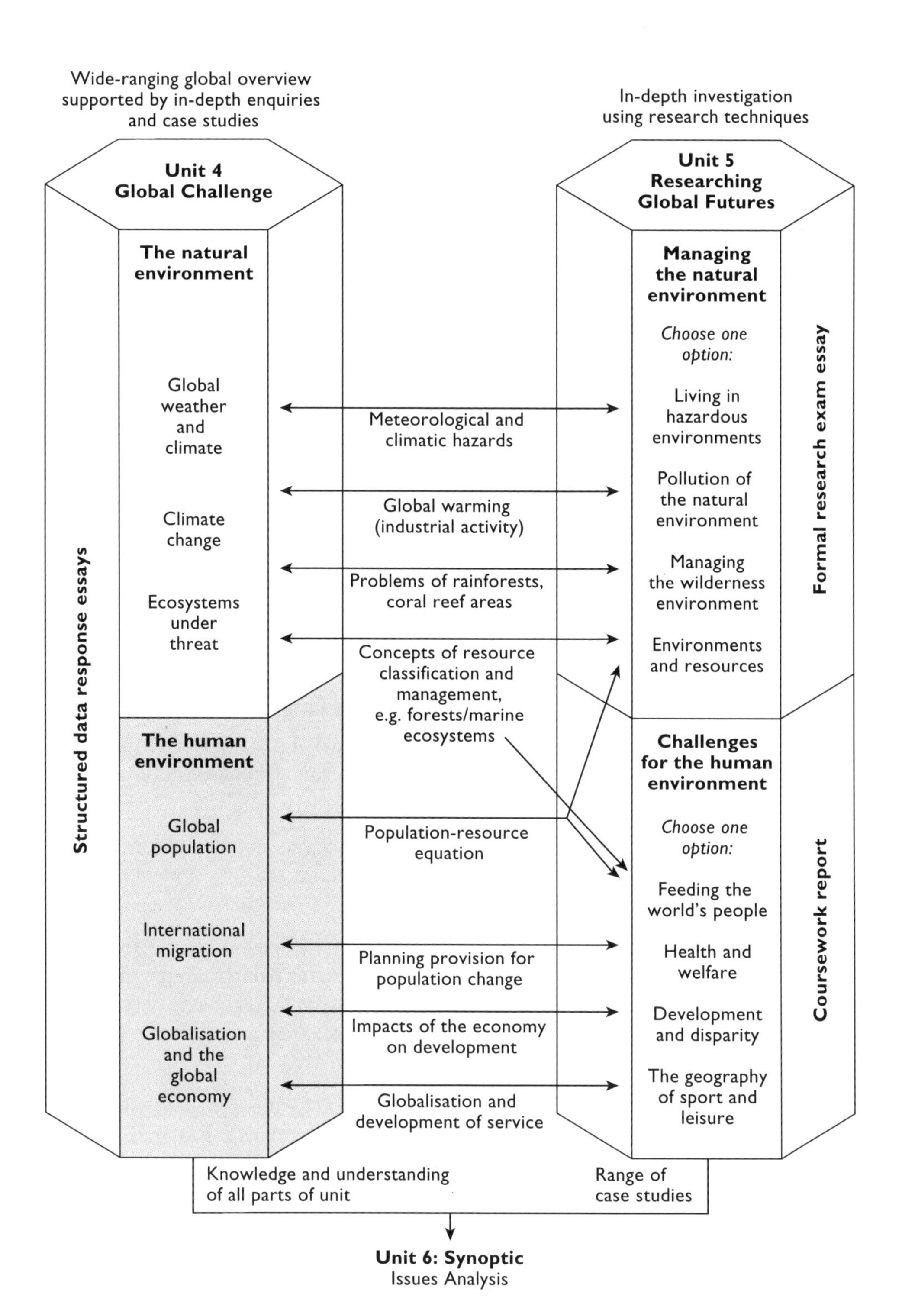
Wide-ranging global overview supported by in-depth enquiries and case studies
In-depth investigation using research techniques
Unit 4 Global Challenge
Unit 5 Researching Global Futures
The natural environment
Global weather and climate
Climate change
Ecosystems under threat
The human environment
Global population
International migration
Globalisation and the global economy
Structured data response essays
Managing the natural environment
Choose one option:
Living in hazardous environments
Pollution of the natural environment
Managing the wilderness environment
Environments and resources
Challenges for the human environment
Choose one option:
Feeding the world's people
Health and welfare
Development and disparity
The geography of sport and leisure
Formal research exam essay
Coursework report
Meteorological and climatic hazards
Global warming (industrial activity)
Problems of rainforests, coral reef areas
Concepts of resource classification and management, e.g. forests/marine ecosystems
Population-resource equation
Planning provision for population change
Impacts of the economy on development
Globalisation and development of service
Knowledge and understanding of all parts of unit
Range of case studies
Unit 6: Synoptic
Issues Analysis

The knowledge and understanding gained in this unit will also be partially assessed in Unit 6, a synoptic issues analysis paper. Unit 6 will test your ability to draw together what you have learnt and understood, drawing on the skills that you have acquired in the context of a particular issue in an unfamiliar location or context. Issues analysis draws from all the units.

The unit test

Timing

The examination lasts for 2 hours and counts for 15% of your A2 mark. The questions in Sections A and B are each worth 25 marks whereas Section C is worth 30 marks. Therefore, you are advised to spend more time on your answer to Section C — approximately 45 minutes. You should spend 35 minutes each on Sections A and B.

Divide up your time between the three sections. Either stick to a rigid 35 minutes for A and B or, alternatively, do Section C first if you see a question that you can do. People frequently overrun on their first question, so you might find it easier to complete the higher-marked Section C question in the first 45 minutes.

Section A is covered in the companion guide, *Unit 4: Global Challenge (The Natural Environment).*

Concepts, theories and geographical terms

At A2, the skill of using the correct vocabulary is essential. It is a good idea to compile a list of key terms as you meet them. Use your textbooks and your class notes to build up your own dictionary.

Key ideas and theories are more important at A2 than at AS, and they will be identified for you in this book using **bold type**. These are often conflicting, and you should appreciate these differences.

This unit assumes that you have left the simplified global division into **LEDCs** and **MEDCs** behind. You should see the state of development as being a continuum from the least developed **LDCs** to **LEDCs**, **NICs**, former Soviet states (sometimes called **transitional economies**), **developed economies**, **MEDCs** and even the leading economies of the **G7**.

This unit is about broad issues that the world faces. The diagram opposite summarises some of the key ideas. Because this is an A2 unit, you are expected to read around these major issues and be aware of the concepts and theories that underpin them.

Global Challenge is a very big unit. Undoubtedly you will become more interested in some issues than others, but because of the way that the examination paper is organised, you are strongly advised *not* to exclude any element from your studies. For instance, if you leave out climate or economy, you could struggle on the cross-unit questions in Section C.

The topics are often current issues in the world news, such as asylum seekers, global warming, threats to forests or grasslands or coral reefs, the debt crisis and the global economy. You can help yourself to succeed by reading quality newspapers and articles in magazines such as *Geography Review*, *New Scientist* and *The Economist*. However, you still need to read textbooks and should not rely on one book.

Examination technique

The main skill that you need is that of writing semi-structured and open-ended essays. Questions are normally in two parts, with the occasional three-part question. The examination paper is accompanied by a resources booklet. This contains a wide variety of resources for you to interpret and use as a stimulus for your answers. These include:

- satellite, vertical or oblique aerial and ground-level photographs
- maps at various scales
- articles and cartoons from newspapers
- tables of data
- synoptic charts — weather maps

Data require close analysis because they often contain clues to your answer. The skills that you require are summarised below:

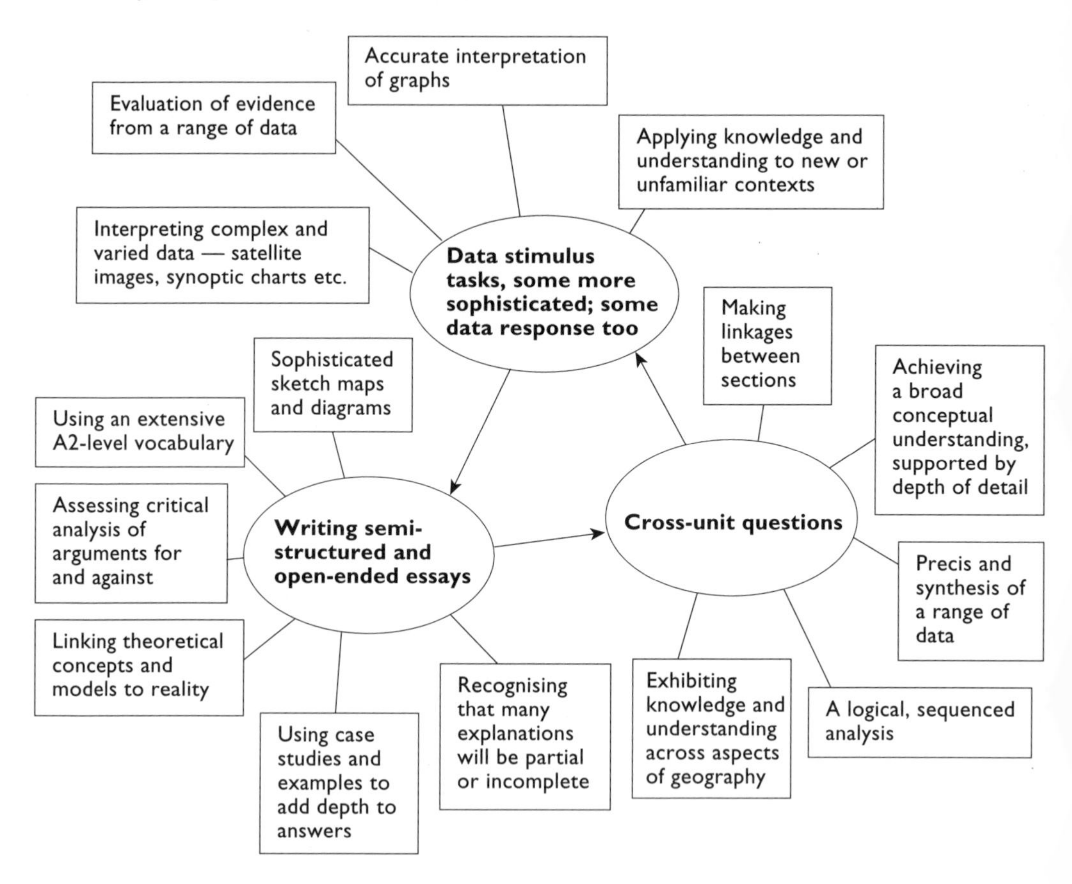

Choice of questions

You must answer *three* questions, one from each section (each containing three questions).

- **Section A: The Natural Environment**. The examiners will ensure that at least one question covers weather and climate and another is on global biomes and biodiversity.
- **Section B: Population and the Economy**. There will always be a question on population and migration and one on the global economy.
- **Section C: Cross-unit** questions.

The examiners will try to ensure coverage right across the specification. If a topic is not included in Section A or B, it is probable that it will be a part of Section C. Do remember that the same topic cannot be set every year and it is most probable that the topics set in the previous examination will not reappear.

There are three stages to selecting questions:

Stage 1
Make sure that you understand the resource and do not be seduced by it!

Stage 2
Look carefully at the knowledge and understanding required and check that you can illustrate your work with appropriate examples and, where required, an in-depth case study. In your first 5 minutes, read through the paper and list your learnt case studies on the question paper so that you know which questions are the best choices. Selecting questions from Sections A and B is normally easier, so it is probably a good idea to attempt these sections first, but *following the strict timing framework suggested on page 6*. The framework does allow more time for planning and developing the answer to Section C, which carries more marks.

Stage 3
Draw up a simple plan for each sub-section that reflects the subject matter of the question, paying attention to the command words.

Managing questions

Command words

Your question management will need to be more sophisticated than at AS. The resources are there to stimulate your responses. Describing a map, diagram or graph will not be enough because the questions will probably ask you to explain or give reasons for what is shown.

Some command words are more demanding at A2. You will often be asked 'To what extent', 'Critically assess', 'Evaluate', and 'Assess the factors that'. These are evaluative questions which anticipate that there is more than one explanation. You are expected to say which factors or explanations are more important and why. You will also be asked to 'Discuss the assertion', a command that expects a discussion of both sides of a case. More questions will expect you to 'Explain', for which description is not enough. 'Analyse' often implies some numerical analysis.

Use of diagrams and maps

It is always worth learning relevant and easy-to-draw diagrams. In this guide, there are many diagrams which you will find very useful, for example when looking at the spectrum of development. Maps and diagrams can save words provided that you integrate them into your answer and summarise the key points. Summary tables may also be used and can be effective.

Time management

Do remember to manage your time (see page 6). Missing a section of a question will probably lose you one grade on the paper.

Only spend a limited time planning. Allow a maximum of 10 minutes for question selection and planning time (5 minutes per question) for Sections A and B. Section C has more time allocated but here it is expected that you do need around 8–10 minutes of thinking time. Although the questions are each worth 30 marks, this reflects the challenge of Section C questions; you will not necessarily have to write more.

Approximate guide to the time and length of answers

Mark	Time in minutes (Sections A and B)	Length in words (approx.)	Time in minutes (Section C)	Length in words (approx.)
18/20	25	700	30	800
15	18	600	20	650
12/13	17	500	15	500
10	12	350	10	350
5	5	200	5	200
Planning time	Up to 5 minutes per question		8–10 minutes	

Note: relevant diagrams and maps may take up to 20% off the time and length figures.

Quality of written communication

Four marks on this paper are added at the end for the quality of written communication. These are extra marks and are awarded for:

- the structure and ordering of your response into a logical answer to the question, using appropriate paragraphing (single-paragraph answers will not score the maximum)
- the appropriate use of geographical terminology
- your punctuation and grammar
- the quality of your spelling (if you are dyslexic, seek special consideration)
- introducing and concluding the sections; these should not be too long, because a quality main body to an answer is required

Content Guidance

The **Population and the Economy** sub-unit of Unit 4 comprises six sections:

- **the dynamics of population change**
- **the implications of population change**
- **the global challenge of migration**
- **introducing the global economy**
- **globalisation and changing economic activity**
- **economic futures**

These six sections form the basis of Section B of Unit 4, but the concepts, theories and ideas that are the backbone of **Population and the Economy** are equally applicable to the cross-unit questions (Section C).

Throughout the Content Guidance, key terms are in **bold**. This should help you to build up your own dictionary of essential terms.

Whereas at AS you were expected to refer to MEDC and LEDC examples, at A2 you are expected to take examples from many of the national economic types shown in Table 6 on page 42.

The dynamics of population change

Population growth and changing populations are among our greatest challenges. The six billionth person was born in October 1999. **Systems theory** provides a framework for studying population because it enables us to see any population as dynamic. **Demography** is the study of population's vital statistics. This section looks at the tools you need to demonstrate your understanding of the fundamental principles.

Population change can be measured at a variety of scales. The world's population doubled between 1960 and 1999. However, the rate of increase varies (Figure 1). The UK's population will take 433 years to double compared with 27 years in Africa. Since 1950 the 48 least developed countries have contributed 12.5% of the world's population growth whereas MEDCs have contributed only 3%. As a result, the distribution of the world's population is changing. World population growth is the first global challenge, especially as it is occurring in those countries least equipped to cope. However, do bear in mind that the art of forecasting is very imprecise and is based mainly on projecting existing trends.

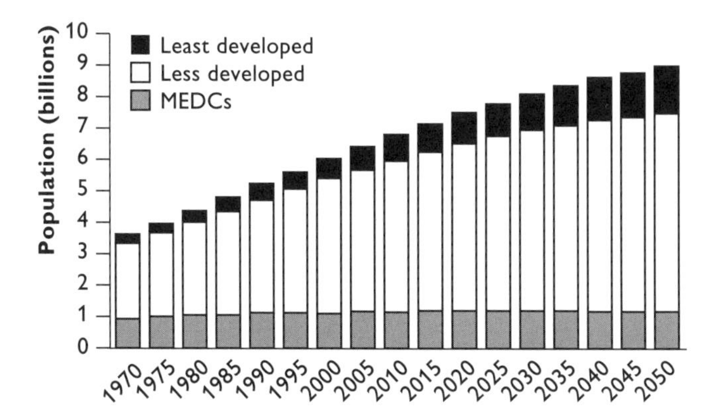

Figure 1 Population growth by development category, 1970–2050

There are now 23 countries with over 50 million people (compared with nine in 1950). In 2000, China had approximately 1250 million inhabitants and India 975 million. The countries with the highest increases at the turn of the century were Islamic ones. In these 40 countries, the total population trebled between 1950 and 1998 so that now 43% of their populations are under 15 years old.

According to the 1991 Census, the population of England and Wales was 49.89 million. As a part of your AS studies, you will have looked at the changing population of an urban and an adjoining rural area. Figure 2 (overleaf) shows the key features: **density**, 1991; **change** 1981–91; **natural increase**, 1981–91; and **net migration**, 1981–91. At A2, you should be able to describe and explain these patterns.

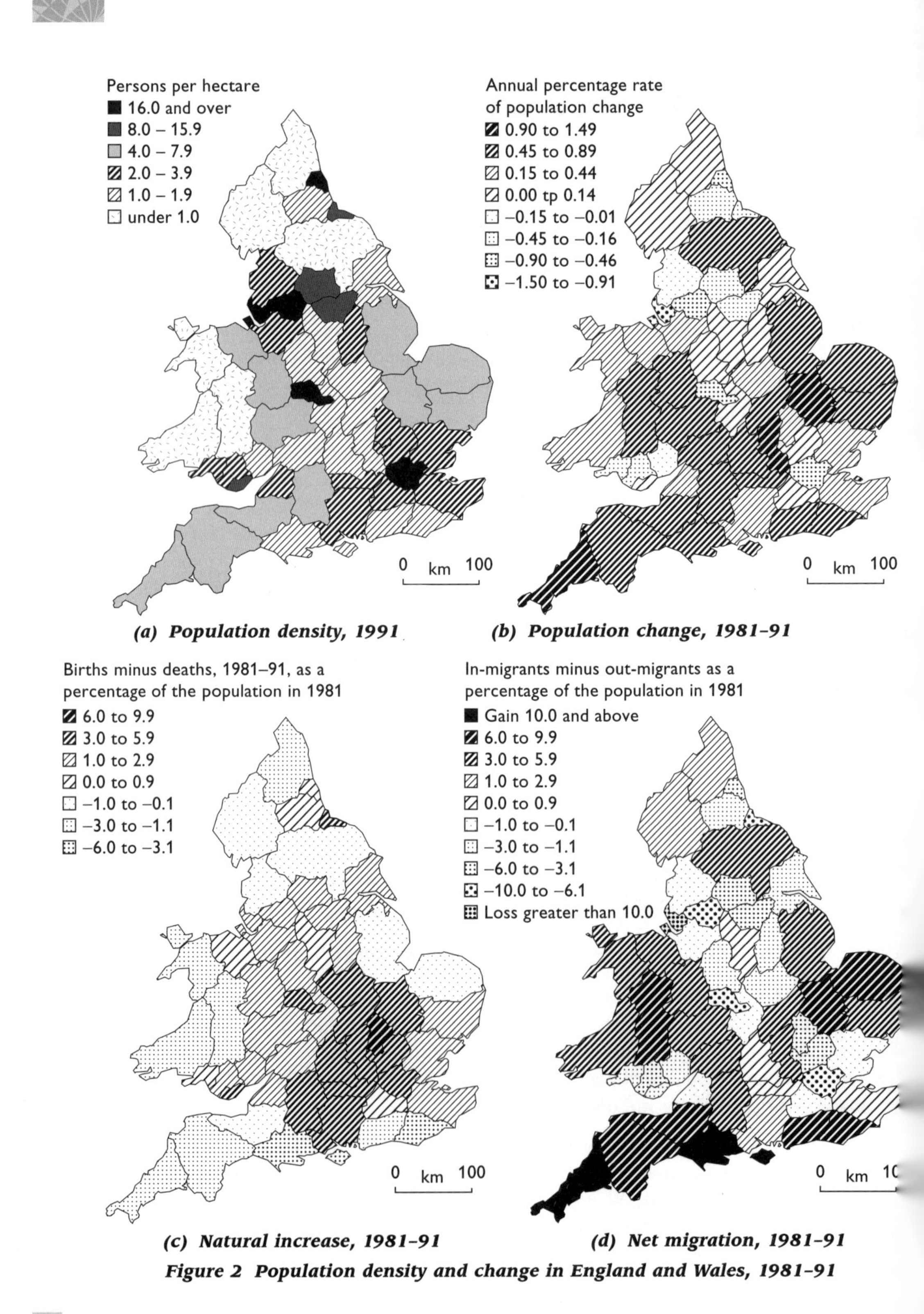

(a) Population density, 1991

(b) Population change, 1981–91

(c) Natural increase, 1981–91

(d) Net migration, 1981–91

Figure 2 Population density and change in England and Wales, 1981–91

Population and demographic data

Population data are normally collected by ten-yearly national censuses. These give us the absolute numbers at that time, either where people are, or where they normally are. Some data are sampled; normally, 10% of households is the sample size.

Key terms

Natural increase is the excess of births over deaths.

Crude birth rate is the ratio of the number of births per annum to the total population, expressed per 1000.

General fertility rate is generally measured as the number of births per 1000 women aged 15–45 (49 in some cases). This can be refined further to examine the number of births to a specified age group per 1000 women. In MEDCs, this rate needs to be 2.1 children per woman aged 15–45 to maintain the level of the population. In Italy, the rate was 1.2 in 1998, while in the UK it was 1.7. In LEDCs in 2000, the rate was 5.6 children per woman and in Islamic countries it was 4.1. Birth rates and fertility rates can be driven by:

- culture, for example religious beliefs and dogma
- political pressures to increase or decrease the population
- social norms, for example later marriage
- events such as the end of a war

Crude death rate is the ratio of the number of deaths per annum to the total population, expressed per 1000 people. It is possible to refine this to look at the **age-specific mortality rate**. Hygiene, diet and medical advances all affect the rise and fall of these rates. One of the key rates is the **infant mortality rate**, which measures the number of deaths of children aged under 1 year old per 1000 live births. The **maternal mortality rate** measures the deaths of women in childbirth per 100 000 live births. Both of the last two measures are used frequently to measure the development of a country.

Life expectancy is the average number of years a person is expected to live. It can be refined by age groups and gender. In the UK, it is 76 (MEDC average 77.8), whereas in Rwanda it is 39 (LEDC average 59, least developed average 51).

Migration is a movement of people from one administrative area to another which results in a change of permanent residence.

Migration balance is the excess of in-migration over out-migration, or vice versa.

In-migration is the flow of people over a given time period (year or decade) into a country/area, on either a temporary or a permanent basis. **Out-migration** is the opposite flow.

Net population change is the change in people in a country over a period of time when both natural change and migration change have been taken into account. It is normally expressed as a number or a percentage. Figure 3 shows the data for 1995–2000.

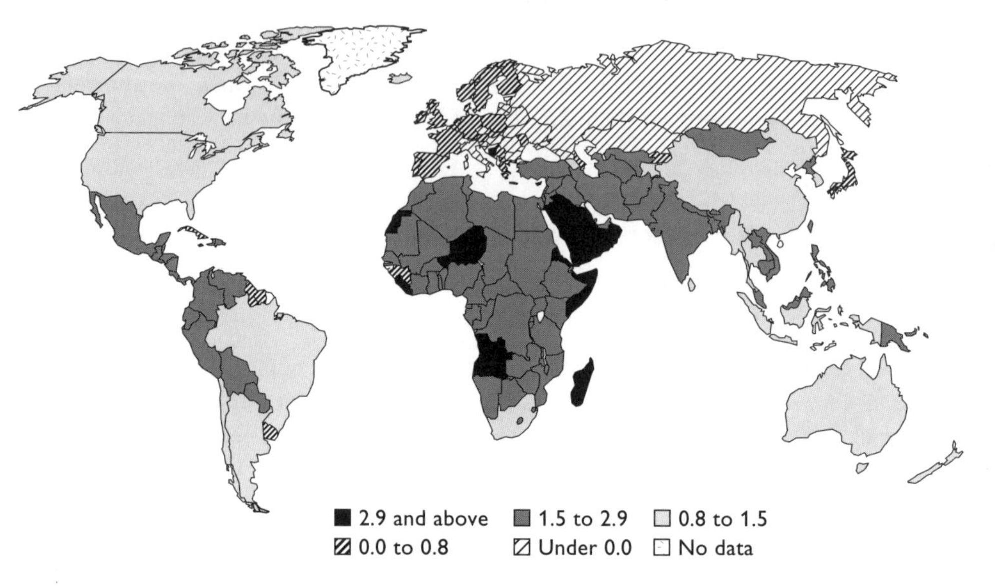

Figure 3 Annual average rate of population change, 1995–2000 (%)

Displaying demographic and population data

Geographers and demographers portray population in a variety of ways. Each one of the measures above can be mapped at almost any scale from the global to the local (city) scale.

Density maps show the number of people per unit of land area (persons per hectare). This is shown as a **choropleth map** (shaded or coloured by areas of differing densities). This technique is used to illustrate most of the above data sources at global and national scales. Density maps ignore the ability of the land to support the population. **Physiological density** measures the number of people supported per hectare of arable land, thus measuring a country's ability to support itself from within its own boundaries.

Distribution is the location of people portrayed by dots or proportional symbols. It is relatively unsatisfactory because symbols overlap, but it can prove useful at a local scale where there is more room to portray the data.

Topological maps portray the global population challenge graphically. The area of each country is in proportion to its total population or other total being measured (Figure 4).

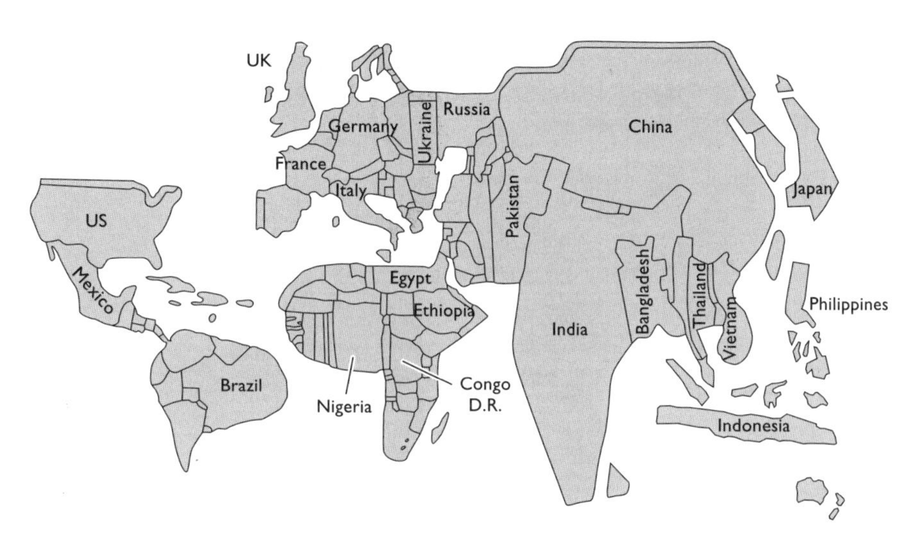

Figure 4 Topological map of the world's population

Migration may be portrayed using the techniques above. It is best portrayed on **flow maps** showing the direction and scale of movements by the width of the flow line.

The **age–sex pyramid** classically portrays gender and age. Pyramids are an indicator of development. They are used to study the demographic characteristics of individual groups in an area. Figure 5 shows some pyramids for the residents of the city-state of Singapore. In order to make pyramids comparable, all of the data should be expressed as percentages of the total population and not as absolute numbers.

Figure 5 Age–sex pyramids for the city-state of Singapore, 1990

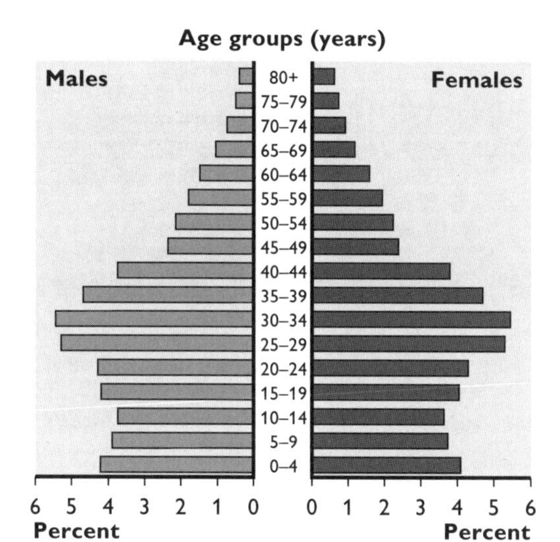

(a) Age–sex pyramid of the total population in Singapore, 1990

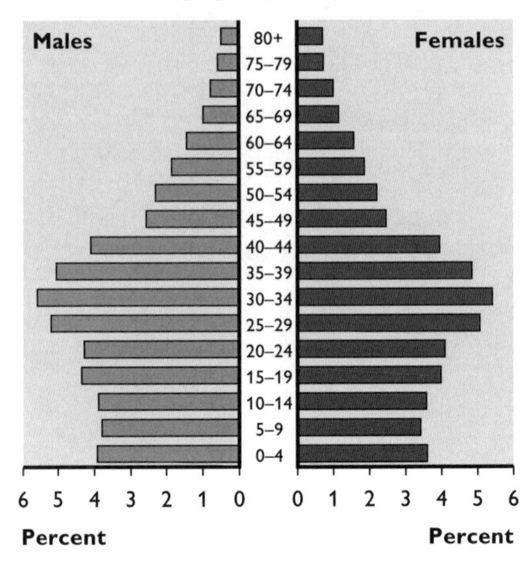

(b) Age–sex pyramid of the resident Chinese population in Singapore

Age groups (years)

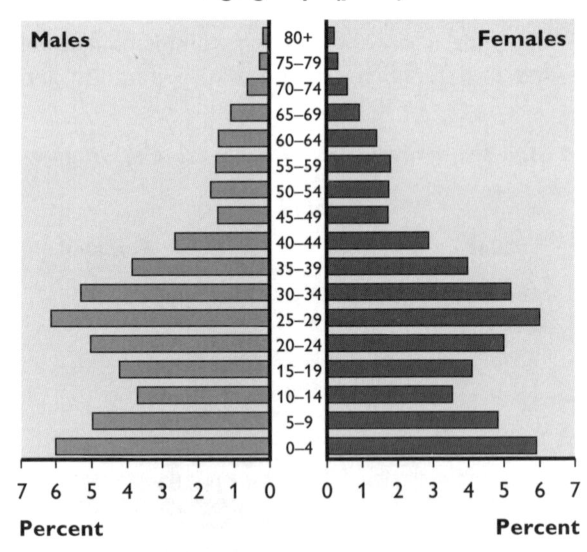

(c) Age–sex pyramid of the resident Malay population in Singapore

Graphs can also be used to show trends. **Line graphs** are used to show changed rates of growth for many of the demographic indicators. **Bar graphs** are often used to illustrate time series for a set of places.

Censuses

Why do we have censuses?

Censuses do not only gather demographic data. They also gather data on employment, education, commuting patterns, internal migration, ethnicity, patterns of social activity and change, and housing type and ownership. These data are used by government and industry. In the UK, the data are plotted by areas of local government, grid squares and postcodes.

At a national level, a census:

- looks at trends over the previous 10 years and projects these forward to enable planning to take place
- projects future demands for jobs
- projects natural change and migration changes
- enables planning of services such as schools, hospitals and services for the elderly
- enables catering for in-migration
- enables the projection of national housing demand
- enables planning for national transport demands
- enables financial and economic planning to take place
- is a snapshot of the diversity of a country

At a regional and local level, a census:

- enables regional plans to be developed to cater for anticipated natural increase/ decrease and migration
- enables services such as schools, libraries and hospitals to be located close to demand
- assists in the provision of essential utilities: water, waste disposal, electricity and gas
- enables space for new housing to be allocated
- locates **multiple deprivation** based on grid squares, postcodes and the small areas covered by the census collector, the **enumeration districts** (ED)

For business and commercial interests, a census:

- is correlated with other data sources, for example credit card data, to provide information on regional lifestyles
- enables targeted marketing using information based on postcodes
- enables the insurance industry better to assess risk
- enables firms to locate in areas where people and required skills are available
- enables retailers to invest in optimum locations where spending power is highest
- enables firms to target goods to stores according to the profile of the population, for example more prepared foods in areas of young, single people

Not everyone wants censuses. They are increasingly opposed as an infringement of privacy — for example, West Germans delayed the census for 6 years between 1980 and 1986. They are also opposed because obtaining the sophisticated data now wanted by business, government and planners costs money. The 2001 UK census did not allocate enough money for enumerators to follow up those who did not submit the completed forms. Consequently, it is possible to question the overall accuracy of the

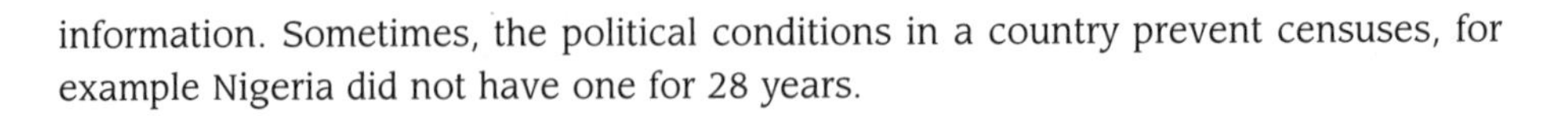

information. Sometimes, the political conditions in a country prevent censuses, for example Nigeria did not have one for 28 years.

Global population change

Table 1 provides some figures for a selection of countries. These countries provide you with a range of data that you can use to illustrate answers.

Table 1 Population data for a selection of countries, 1999

Country	Population (millions)	Life expectancy (years)	Crude death rate (per 000)	Crude birth rate (per 000)	Children per woman	Infant mortality rate (per thousand live births)
Afghanistan	22.8	46	22	50	6.8	n.a.
Vietnam	78.7	68	6	20	2.6	37
India	998.0	63	9	26	3.1	71
China	1253.6	70	7	16	1.8	30
Bangladesh	127.7	58	9	28	3.1	61
Chad	7.5	47	16	45	6.0	101
Sierra Leone	4.9	38	25	45	6.0	168
Burkina Faso	11.6	71	19	44	1.2	105
Burundi	6.7	43	20	41	6.8	105
Rwanda	8.3	41	22	45	6.1	123
Egypt	67.2	67	7	26	3.3	47
Uganda	21.5	40	19	46	7.1	88
Malaysia	22.7	72	4	24	3.1	8
Singapore	4.0	77	5	13	1.7	3
Thailand	60.8	69	7	17	1.7	28
Mexico	97.3	72	5	27	2.7	29
Brazil	167.9	67	7	20	2.3	32
Russian Federation	147.1	67	14	9	1.3	16
Bulgaria	8.2	71	14	8	1.2	14
Germany	82.1	77	10	9	1.3	5
France	58.8	78	9	13	1.7	5
Italy	57.6	78	10	9	1.2	5
Japan	126.6	80	8	10	1.4	4
USA	278.2	77	9	15	2.0	7
UK	59.5	77	11	12	1.7	6

You should be able to draw quick sketches of the following global demographic characteristics from your own textbooks and learning:

- net population change
- crude birth rate
- fertility rate
- infant mortality rate
- death rate
- life expectancy

More important than the actual distributions and figures are the reasons for them. Figure 6 illustrates the key factors explaining global variations in demographic indicators. The main factor is the **stage of development**. This in turn is the product of the factors shown in the diagram. Remember, the indicators of demographic health and all of the data are fixed in time. The reality is that countries and regions are all dynamic and changing. Countries develop both economically and socially, and this affects population change over time.

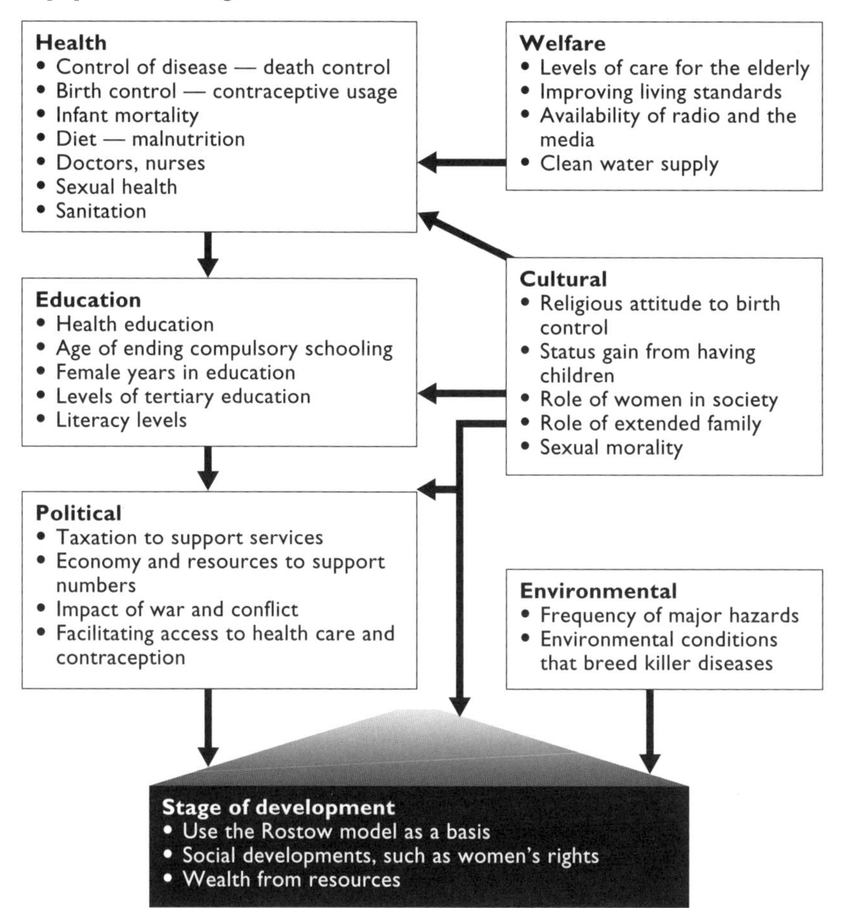

Figure 6 Explaining global variations in demographic indicators

Population change over time

The most common explanation for change is the **demographic transition**, a process of four stages. Countries have distinctive characteristics at each stage. No country is now in Stage 1, low growth. Stages 2–4 are described and illustrated below. The model ignores migration, which is significant for many countries.

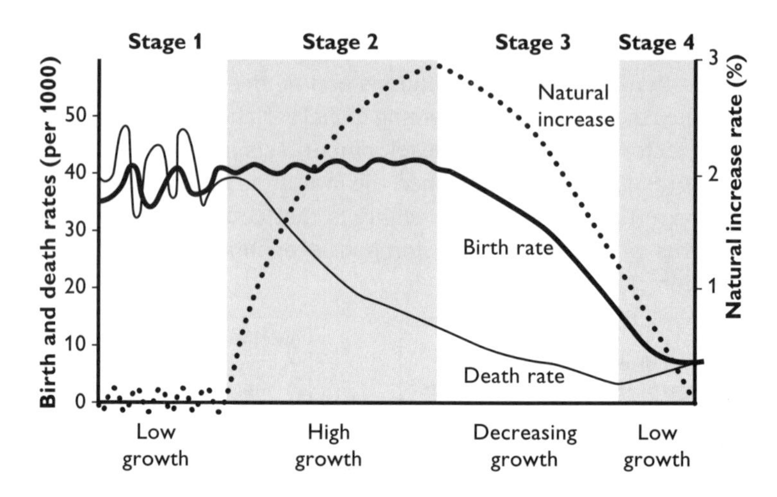

Figure 7 The demographic transition model

Stage 2 (e.g. Cape Verde)

- Accelerating growth with a large gap between birth and death rates.
- Medical revolution enabling death control.
- Low infant mortality.
- Increasing urbanisation and industrialisation.

Figure 8 Population pyramid and demographic transition for Cape Verde

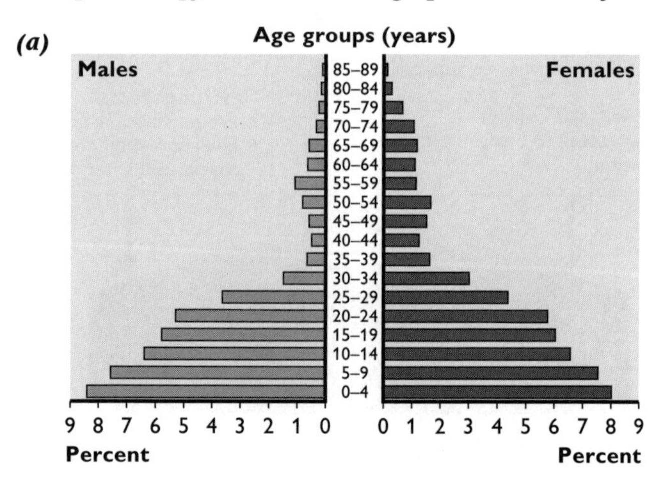

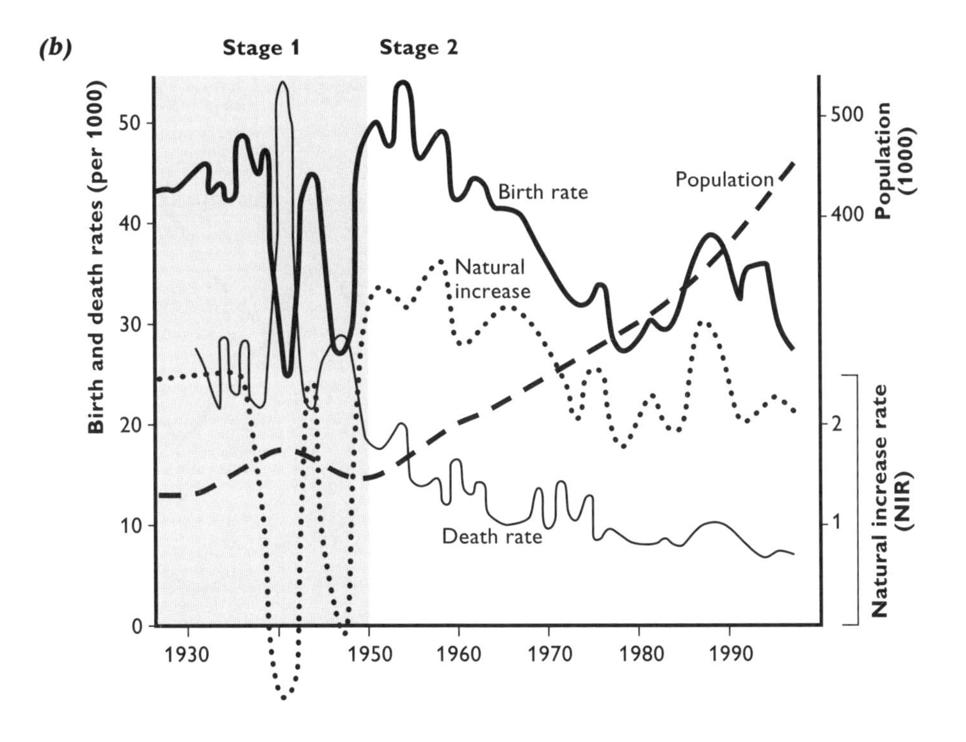

Stage 3 (e.g. Chile)

- Moderate growth because birth rate is falling.
- Social customs change.
- Greater materialism and altruism — concern for life chances.
- Economic changes encourage fewer children.
- Economy diversifies into services.

Figure 9 Population pyramid and demographic transition for Chile

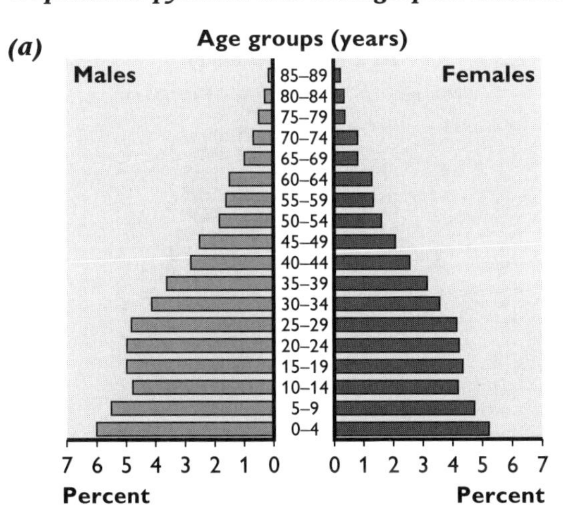

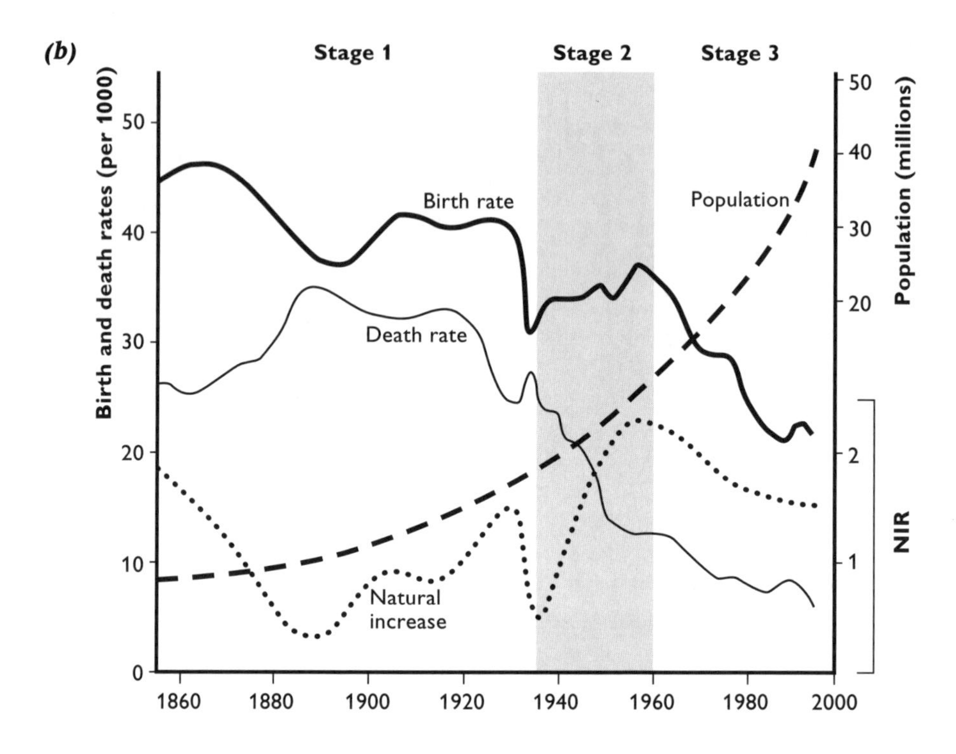

Stage 4 (e.g. Denmark)

- Low, if not zero, growth.
- More women in the labour force.
- Lifestyle changes — higher incomes, more leisure and increased birth control.
- In-migration may keep numbers from total decline.
- Post-industrial society.

Figure 10 Population pyramid and demographic transition for Denmark

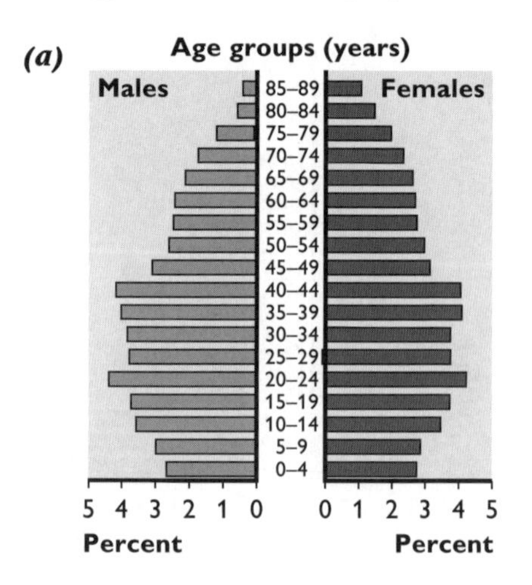

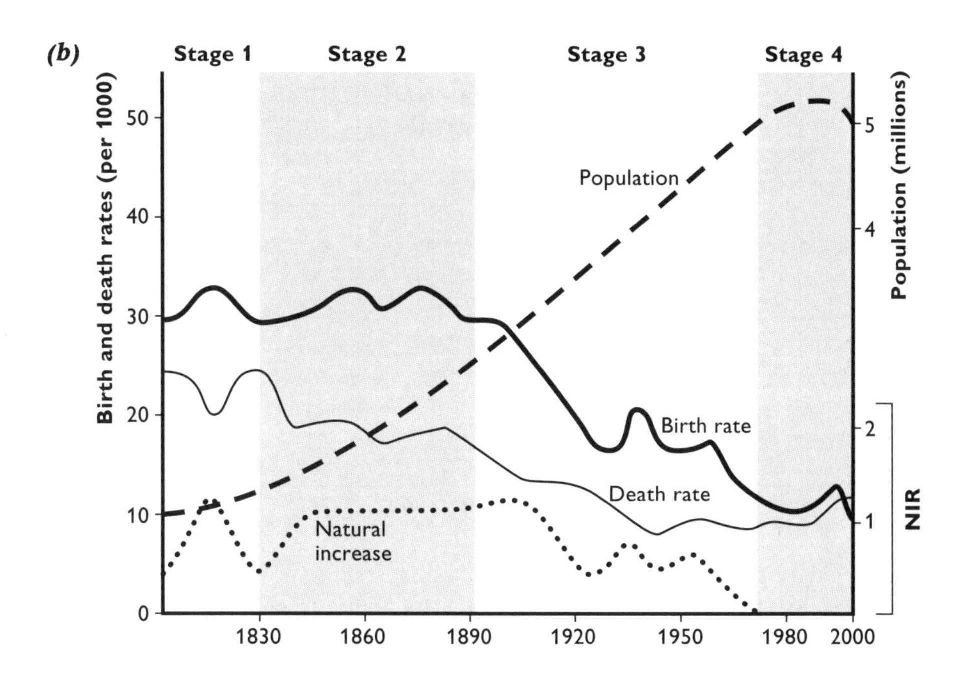

Stage 5 — 'second demographic transition'?

- Noted by Belgian and Dutch demographers.
- It is suggested that there are MEDCs with increasingly old-age populations, low in-migration and high emigration of young, qualified people.
- Economy is information- and ICT-based.
- Rise of individualism linked to emancipation of women in the labour market, new attitudes to contraception and abortion, and greater financial independence.
- Greater environmental concerns for the impact of increased numbers on the resources for future generations.
- Rise of non-traditional lifestyles, such as same-sex relationships.
- Rise of childlessness and delayed pregnancy.

The implications of population change

World population is growing **exponentially** or **geometrically** — the number increases as a constant proportion of the number at a previous time. It appears as an increasingly steep curve on normal graph paper and a straight line on logarithmic graph paper. Table 2 (overleaf) tabulates our concerns by continent, using forecast data for 2050.

Table 2 Population growth, 1950–2050

	Population (millions)			Share of world population (%)			Change (millions)		Share of world change (%)	
Continent	**1950**	**1998**	**2050**	**1950**	**1998**	**2050**	**1950–98**	**1998–2050**	**1950–98**	**1998–2050**
Africa	221	749	1766	8.8	12.7	19.8	528	1017	15.6	33.9
Asia	1402	3585	5268	55.6	60.8	59.2	2183	1683	64.6	56.1
Europe	547	729	628	21.7	12.3	7.0	182	–101	5.4	–3.6
Latin America	167	504	809	6.6	8.5	9.1	337	305	10.0	10.2
North America	172	305	392	6.8	5.2	4.4	133	87	3.9	2.9
Oceania	13	30	46	0.5	0.5	0.5	17	16	0.5	0.5
World	**2522**	**5902**	**8909**	**100**	**100**	**100**	**3380**	**3007**	**100**	**100**

To cope with population growth there have to be resources at the global, continental or national level to support the population. The ability to support a population is measured by **population–resource ratios**, a dated attempt to link the ability of a population to support itself from its own resources and by trading.

Sustainable population, once called **optimum population**, is the population that can be supported so that natural resources are not depleted and output is maximised within the prevailing technological, economic and social conditions. It is a theoretical aim for the world and more so for many individual countries.

The major concern for the future is **overpopulation**. This is where resources are unable to sustain a population at the existing living standard without a reduction in that population or an increase in the resources available to support it. It is a concern for LEDCs in particular, where Malthusian checks such as famine have an impact (e.g. Sahelian countries).

Underpopulation is rarely a concern today, although it might become one in many MEDCs, where populations show signs of natural decline. International migration is often encouraged in these circumstances. It occurs when existing resources could sustain a larger population without lowering living standards, or where the population is too small to develop the resources available.

The debate over the ability of the world to support its population has been around since Thomas Malthus wrote his theory at the end of the eighteenth century. In the twentieth century, Esther Boserup and the Club of Rome respectively challenged and supported Malthus, as demonstrated by Figure 11. Today, we are more aware than ever of the finite nature of the world's resources. This has consequently increased the importance of **sustainable development** and the critical links between the population, the economy and the environment.

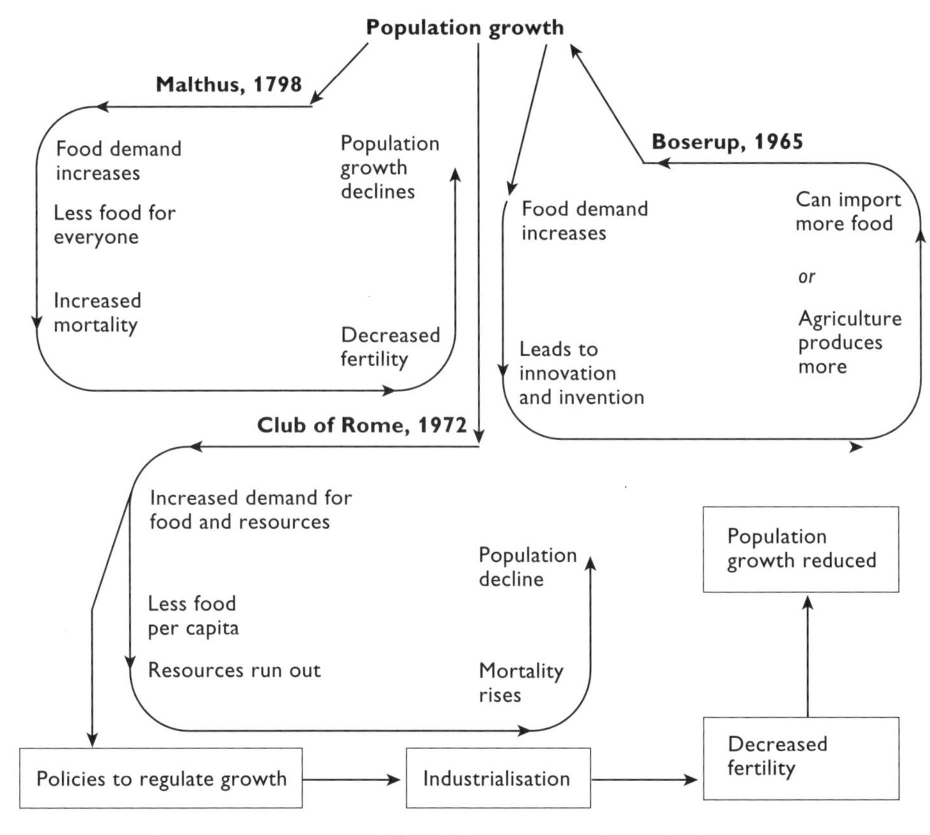

Figure 11 Three models of the impact of population growth

Coping with population change

Growth in a confined space

Singapore

Singapore has a population of 4 million people in a small area, with no natural resources other then the ability of the people. Singapore's policy is to encourage parents to have only one child. To provide the wealth necessary to support the growing population and to improve living standards, the state has invested in high-technology industries and in creating a financial centre for the region. It has built a science park and a major conference centre.

Singapore encourages in-migration of skilled labour from Europe to supplement the local workforce and enhance the profitability of the service sector. These activities earn money to support the building of new towns, for example Tampines, to house the population. More recently, the state has sponsored industrial investments overseas in Indonesia (Riau Islands), Malaysia (Johor) and China to gain the profits of being the regional headquarters for overseas workbench assembly lines. Low-skilled, low-profit activities have been relocated out of the country so that high-skilled, high-profit activities can take place in Singapore.

The Singapore government has invested heavily in education and has encouraged its educational system to invest overseas. It has developed a global airline to provide foreign earnings from its activities (Singapore Airlines holds shares in Virgin and Air New Zealand). All of these efforts are underpinned by strong state planning with five-year plans. As a result, Singapore has the monetary resources to support and sustain moderate population growth.

Growth in LEDCs

India

India has made strides to reduce its population, yet it is still 115 out of 162 countries on the UN Human Development Index (HDI). The birth rate was 41 per 1000 in 1951 and had fallen to 26 by 1999. At the same time, the death rate declined from 23 to 9. In 1991, the average household size was 5.52 people.

In 1998, 36% of the population was below the **poverty line**, calculated as being able to receive 2400 calories per day in rural areas and 2100 calories per day in urban areas. Kerala has only 10% poverty, whereas Bihar and Rajasthan have over 50%.

Family planning, now called **family welfare**, is encouraged, as is sterilisation (at its peak in 1976, 8.3 million sterilisations were performed). Abortion has been permitted since 1972. In addition, **feticide**, the abortion of girl fetuses, is common among middle and upper classes in the richer areas such as Delhi. Female infanticide, because of dowry problems, is still practised in Rajasthan. Therefore, India has a **gender imbalance** irrespective of policies to curb population (882 women for 1000 men in 1995). Education, particularly of women (illiteracy is 65% among women and 36% among men), is forecast to assist in lowering the birth rate — for example, travelling contraceptive clinics are being introduced in rural areas.

Other population policies have focused on feeding the school population (60 million children under 4 are undernourished) and building rural homes to raise standards of living. Grain production quadrupled between 1951 and 1995 thanks to the **green revolution** and better health control. Improved health provision has halved infant mortality.

Declining population in MEDCs

The UK

The UK no longer has a classic population pyramid — it is top-heavy due to low birth and death rates. The contraceptive pill (post-1963) resulted in 1.7 children per woman by 1999. One in five conceptions do not lead to pregnancy due to abortions. Net international immigration helped to counteract the losses due to the low birth rate.

Average household size was 2.4 persons, with 26% in single-person households, in 1996. Single-person households have increased because of:

- more elderly loners due to longevity
- divorce, with men leaving the parenting unit
- single motherhood
- more affluence, enabling single life
- young adults moving away for higher education and their first job

Consequences are serious for housing plans if the singles trend continues — there will be 4.4 million extra households by 2016. Demand will not be spread evenly, being higher in the south and in rural areas.

There are more elderly people in the population — 10 million in 1994. In that year, 37% of women and 12% of men aged 65–75 were widowed. This rises to 32% of men and 64% of women aged over 75. There is, therefore, more demand for facilities to cater for the elderly, such as sheltered housing, nursing homes and warden-assisted homes. These have to be paid for from national and local taxes, savings or insurance policies. Increasing numbers of single elderly also cause 'bed-blocking' — unintended convalescence in hospitals when there are no family carers or nursing homes available. This adds to the costs of hospitals. Figure 12 shows projected proportions of the elderly population in the UK.

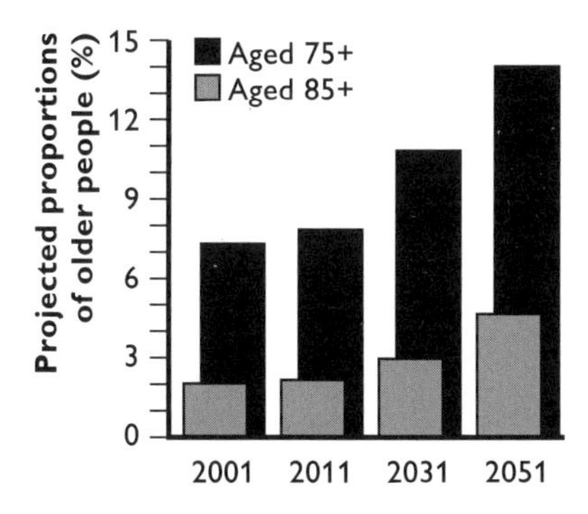

Figure 12 Projections of elderly population in the UK, 2001–51

Decline in LEDCs

Botswana

Botswana is a small, landlocked, semi-desert country in southern Africa with a population of 1.6 million. It has a maternal mortality rate of 330 per 100 000 live births, under-5 mortality of 48 per 1000 and one doctor for every 4762 inhabitants. Its population would grow more rapidly but for the ravages of HIV/AIDS. Life expectancy is 47 years, but this is falling because of AIDS-related deaths. It is estimated that by 2005 life expectancy will be 41 years. The United Nations Development Programme suggests that this will fall to 33 years by 2010. The population is projected to be 20% smaller by 2015 than it would have been in the absence of AIDS. In 1999, 36% of adults were infected with the HIV virus. The spread is linked closely to poverty, particularly because the victims lack education, information and access to health care. Women are more vulnerable if they are poor and in a society tolerant of extra-marital sex, where their partners are promiscuous.

Uganda

Uganda had a population of 21.1 million people in 1999. Growth has slowed for a variety of reasons. There is a maternal mortality rate of 510 per 100 000 live births. The mortality rate for under-5s is 134 per 1000. Illiteracy stands at 38% and there is only one doctor for every 25 000 inhabitants. HIV/AIDS has had a similar effect to that in Botswana. The government has recognised that the issue is developmental and that it will take time to change attitudes. It has begun campaigning against high-risk behaviour, empowering communities through the work of NGOs, educating employers to assist, and providing better access to health care. Today, only 8.3% of adults have HIV, because of the success of these measures over the past decade.

Changing age structures

Figure 13 shows how age structures have changed in MEDCs and LEDCs.

Impacts and issues in MEDCs

- In 1998, the over-60s outnumbered the under-15s for the first time.
- The proportion of elderly people is rising fastest in Europe — 33% by 2050.
- In 2050, Spain will have 3.6 persons over 60 for every one under 15.
- The cost of funding of pensions and care may result in changes to pensionable age.
- Fewer people will be moving into work.
- There will be more employment for the elderly (e.g. Tesco and Homebase).
- There will be less demand for schools and teachers.
- Retailing changes will be needed to service the new age structure.
- Leisure facilities for the elderly (e.g. Saga holidays and tea dances) will increase.
- Fewer people in work will have to support the dependent population.
- There will be less demand for goods and services from the smaller working population.
- In-migration will need to provide workers and to retain wealth-generating ability.
- There will be retirement migration — normally followed by return migration in very old age.

Impacts of ageing in LEDCs

- By 2050 there will be a similar proportion of over-60s in LEDCs as in MEDCs, but within a larger total population.
- The percentage of the elderly in the population will not change because the birth rate will still be high.
- There will be migration of the working population and particularly those with skills to MEDCs.
- Costs of care will rise if there are no relatives to act as carers.

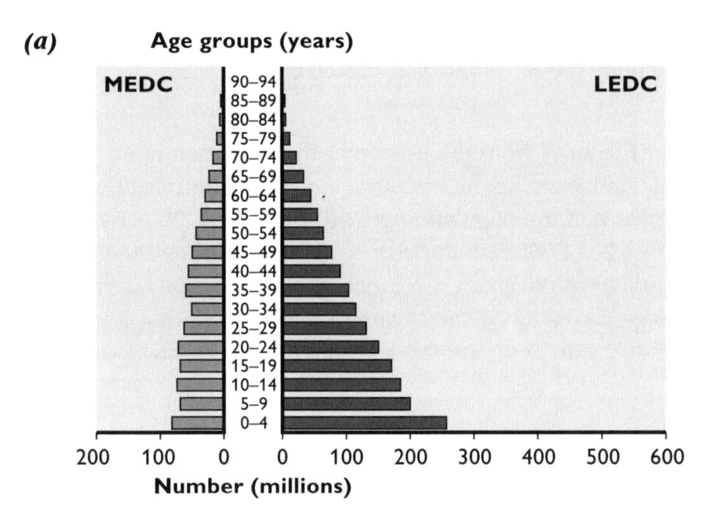

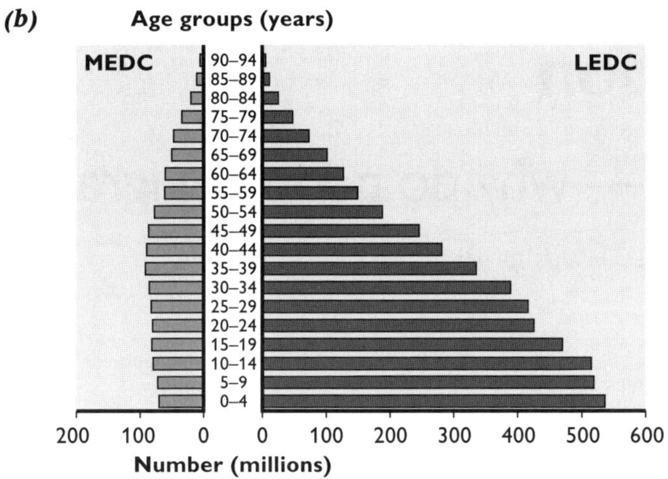

Figure 13 Age structure of MEDCs and LEDCs, (a) 1950 and (b) 2000

National policies — decision makers and population change

Thailand

In 1969, women averaged 6.5 children each, 16% used contraception and population growth was 3% a year. The solution was a nationwide family planning programme which commenced in 1970 and included free contraception, trained family planning specialists and government campaigns, especially among rural communities. By 1999, contraceptives were being used by 72% of people, women averaged 1.7 children and population growth was only 0.8% a year. The population will double by 2084. It is a successful community-based rather than coercive policy.

Philippines
Opposition to birth control from the Roman Catholic Church (83% of the population are Catholic) affected government encouragement of contraceptives. In 1999, 47% used contraceptives and the population growth rate was 1.7% per year. The population will double by 2027. Women average 3.6 children each. The government is an enthusiastic supporter of the green revolution, increasing agricultural output. Labour out-migration (e.g. nurses to UK, musical groups playing in the resorts of Malaysia, and servants, child carers and nannies for the wealthy in Singapore) has not been discouraged.

The global challenge of migration

Motives — why do people migrate?

Figure 14 provides a basic model of migration at any scale.

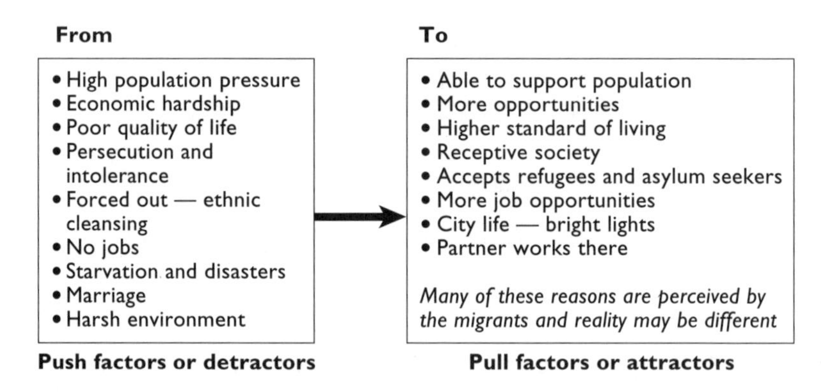

Figure 14 Motives for migration. Do you have examples of these motives at an international scale?

Duration

Unit 4 focuses on international migration. This is generally long term and mostly for economic reasons. However, much publicity has been given to refugees and asylum seekers.

	Short term					Medium term						Long term
	Movement					Migration						
Type	Leisure	Commuting	Weekend	Holiday	University	Job moves	House moves	Expatriot	Retirement	Refugee	Asylum	Emigration
Timescale	Part-day	Day	2 days	2 weeks	3 years	Several years				Temporary to life		Life
Typical	Normally within a country					Possibly international		International				

Figure 15 Duration of migration

Voluntary or forced migration?

Most migration is the result of free choices made by a migrant, the head of a migrant household, or a group. However, according to the UNHCR, 22 million people have currently felt or been forced to leave their homes and/or country. Examples of voluntary and forced migration are given in Table 3.

Table 3 Types of international migration

Voluntary	Example	Forced	Example
Between MEDCs	Italians to Bedford in the 1950s: brickworks	Between MEDCs	East to west Europe; ethnic Germans from Hungary to Germany in 1945
Skilled labour	Financiers to Singapore and New York	n.a.	n.a.
Between LEDCs	Lesotho to South Africa	Between LEDCs	Rwanda to Congo DR
Labour migrants	Caribbean to the UK in the 1950s; Mexicans to the USA	Labour migrants	Lesotho to South Africa
Refugees	Montserrat to Antigua following the volcanic eruption	Refugees	Rwanda/Burundi to Congo DR; Kosovans to Albania (ethnic cleansing)
Asylum seekers	Albanians to Italy; Afghans to Australia	Asylum seekers	Kurds from Iraq to Italy

Models of migration

Ravenstein's laws

Ravenstein's laws are based on Britain in the 1880s. He gave migration the following characteristics:

- It is short.
- It is step-by-step.
- Longer distances are travelled to major centres.
- There are reverse flows.
- Rural people are more inclined to migrate than urban people.
- Females move more than males.
- Most migrants are adults.
- Large towns grow more by migration than by natural increase.
- Volume grows as industrialisation progresses.
- The major flow is from agriculture to industrial towns.
- The major cause is economic.

Do these 'laws' apply today? If so, where and why? If they do not, why is that so?

Lee's model

According to Lee, people:

- assess and perceive the destination
- assess conditions where they are
- look at obstacles between the two, such as distance and cost
- consider personal circumstances

Stouffer's intervening opportunities model

This model attempts to explain why migrants settle for locations other than their original destinations.

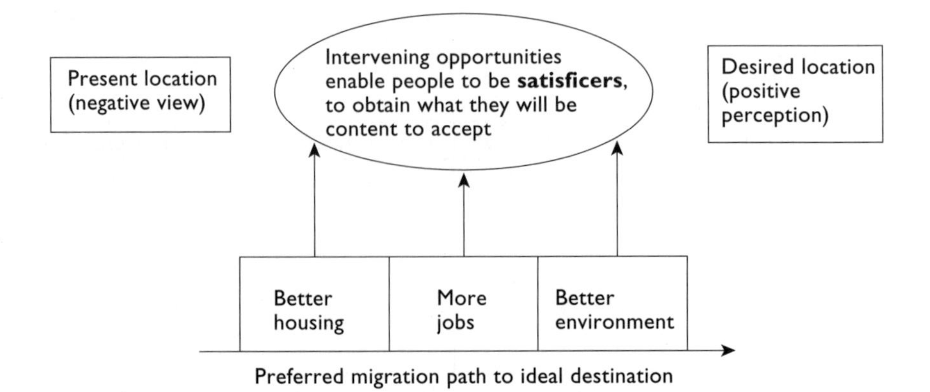

Figure 16 Stouffer's intervening opportunities model

The decision-making model

The decision-making model is shown in Figure 17.

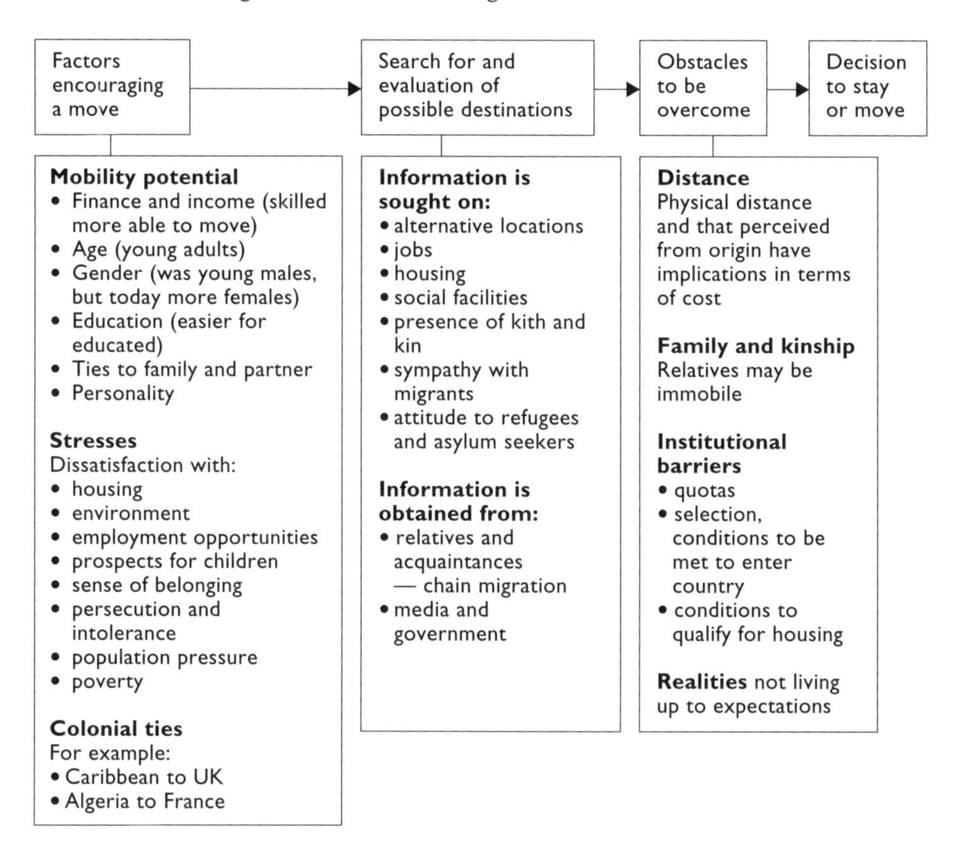

Figure 17 The decision-making model

What is the impact of international migration?

Castles and Miller say this is the **age of migration**. More countries are being affected by international migration. The number of people migrating is growing. There is also more variety among migrants and no type dominates the flows as it did in the past. More women are migrating as labour migrants in their own right. All of these developments pose new challenges for states.

Impact on the country accepting the migrants

- Demographic replacement for low natural change and declining population, for example Turks to West Germany in the 1970s and Italians to Switzerland.

- Labour force needs are satisfied, for example Caribbeans to the UK in the 1950s, Portuguese to Guernsey in the 1990s, Filipino nurses to the contemporary UK Health Service. Some movements are permanent but many comprise **guest workers**: people recruited for a period of time (e.g. Greeks to West Germany in the 1970s) or on assignment (e.g. employees of Barclays Bank working in New York).
- Increased pressure of population on resources, such as food and land (e.g. Rwandans to Zaire), housing and social resources.
- Creation of a **multicultural society**, for example in Toronto. This affects the religious mix, retailing, restaurants, music etc. in the host country.
- Areas become dominated by an immigrant group, for example Turks in Cologne, Italians in Stuttgart in the 1980s, Poles in Illinois, and Caribbeans in New York State and New Jersey.
- Issues of race relations and **segregation**, for example the 2001 report on Oldham and Bradford. **White flight** and **ghettoisation.**
- Transmission of disease, for example flu to Amazonian tribes.
- Gender concentrations, for example male workers from India and Bangladesh to the United Arab Emirates, Mexican women to the USA.
- Illegal labour on low wages, for example Mexicans on farms in California, Thais and Filipinos in the Taiwanese textile industry.

Impact on the country sending the migrants

- Slowing down of natural increase because the fertile migrate.
- Old-age society, as young adults leave.
- Slightly decreased pressure on resources, for example after the Irish potato famine.
- Fewer people to engage in agriculture and produce food.
- Receipt of returned earnings, for example from Dubai to India and Bangladesh.
- Return migration with new skills or prospects, for example taxi drivers and chefs to Mediterranean resorts.
- Westernisation and cultural imperialism reinforced by returnees.
- Solution to political or racial issues, for example Uganda's Asians ousted in the 1970s.
- Loss of skills, for example Indian software experts to USA.
- Pensions outflow where old-age is migration taking place, for example UK pensions being spent in Spain.

Refugees, economic migrants, asylum seekers and ethnic cleansing

It is difficult to distinguish between the following:

- **Economic migrants** — people seeking opportunities.
- **Refugees** — those genuinely fleeing from persecution and harsh political regimes.

A refugee is defined by the UN as:

> a person unable or unwilling to return to his/her homeland for fear of persecution based on reasons of race, religion, ethnicity, membership of a particular social group or political opinion.

- **Illegal immigrants** — people who enter a country without authority and hope to remain.
- **Asylum seekers** — people who request refugee status in another state.

Cases such as Cubans fleeing to the USA during the past 40 years and the Vietnamese boat people in 1975 point to the dilemma, as do the Afghanis found on a freighter near Christmas Island in 2001, trying to reach Australian territory.

Refugees

The UNHCR estimated that there were 50 million people displaced forcibly from either their home areas or their home countries in 1997 and that 14.5 million of these were refugees. Some countries (e.g. Norway, Sweden, Finland and Switzerland) have had a long tradition of accepting refugees and asylum seekers who are unable to stay in their intended country of asylum. Canada adds gender persecution to its definition.

Rwanda and Burundi

In Rwanda the **genocide** of at least 500 000 people as a consequence of civil war between the Tutsi and Hutu peoples in 1994 resulted in the movement of 1.75 million refugee Hutus to neighbouring countries. As Hutus left, a counter-flow of 700 000 Tutsi refugees, many of whom were newborn, began. The UNHCR encouraged repatriation and about 160 000 Hutus returned, but the high birth rate among the refugees actually meant that the population remaining outside Rwanda stayed the same. The refugees were housed in vast camps near the Rwandan border in Congo DR (formerly Zaire) and Tanzania. In 2002, the refugees in Goma were forced to move again by the eruption of Mount Nyiragongo.

In Burundi, a similar tribal conflict, commencing in 1993, was worsened by the flight of Hutus from Rwanda. Many were forced back or fled to Tanzania. The political situation in eastern Congo, where civil war was raging, resulted in the further displacement of 400 000 Rwandans and Burundians across Congo DR and to other states, while 685 000 others were forced by the Congolese to return to Rwanda and Burundi. By 1996, 1.7 million Rwandans and 200 000 Burundians were in camps. At the end of the crisis in 1997, 483 000 persons were repatriated from Tanzania and 234 000 from Congo, and a further 215 000 persons remained unaccounted for.

The crisis resulted not only in the deaths of over a million people, but also in environmental degradation, especially around the sites of the refugee camps where areas were denuded of timber. War and flight halted all development. Many people are still deterred from returning. Therefore, many still live in poverty in camps dependent on aid. Some 2.2 million people still receive aid both inside and beyond Rwanda and Burundi. There is, however, no indication that the troubles will re-ignite and that further refugee flows will follow.

Illegal immigrants

There are an estimated 2.6 million illegal immigrants in western Europe. Nearly all countries are attempting to tighten immigration controls, often in the wake of xenophobic (right-wing nationalist) attacks on refugees and illegals. In 2002, an anti-immigrant politician, Jean-Marie Le Pen, even challenged in the final stage of the French presidential election.

Some people employ illegal immigrants on low wages in order to increase profit. Nearly 300 Poles were returned from Southampton in 2001 after they were found to be working in various activities for minimal wages and in poor living conditions.

Asylum

A definition of asylum is 'the formal application by a refugee to reside in a country when they arrive in that country'. The number of asylum seekers is invariably larger than the number who request refugee status. The numbers seeking asylum have increased steadily since 1970, just when states decided to curtail immigration. One catalyst was the German realisation that Turkish guest workers were unlikely to return home. The collapse of communism in eastern Europe also posed a perceived threat from increased migration which began to be realised in 1992 with large numbers of asylum seekers from the former Yugoslavia.

Reasons for the increased prominence of asylum seeking

- Pressures to migrate from the poorest states are increasing due to a combination of economic decline and political instability.
- Improved communications are enabling people to learn more about potential destinations.
- Communities already in the wealthy states have encouraged others to join them. (**Chain migration** involves members of an extended family following one another.)
- The cost of transport has declined.
- More gangs of human traffickers are preying on would-be migrants and offering passage to a new life.
- The destination countries have found difficulty distinguishing those fleeing from threats to their life and liberty and those trying to escape poverty and improve the quality of their life.

Costs of asylum seeking in destination countries

Increased costs associated with asylum seekers include:

- housing
- social services
- schooling
- welfare payments (e.g. vouchers), all of which must come from taxation
- dispersal (e.g. to Sighthill, Glasgow)
- control policy at points of entry (e.g. the Channel Tunnel)
- policing the right-wing objectors and occasional illegal attempts to deter asylum seekers

Asylum seekers often wish to go to the heartland of a country, for example London or Milan, and not to the depressed areas.

Policy reactions

Policies introduced to deal with asylum seekers include:

- limiting the seekers at source (e.g. more difficult visa requirements)
- pre-boarding arrangements to prevent departure for destinations where asylum may be sought (e.g. the seekers must have a return ticket)
- preventing illegal crossings (e.g. between Cuba and Florida)
- returning ineligible asylum seekers immediately and requiring the carrier to pay for the return
- fast-track procedures to enable entry of genuine seekers
- accepting those with skills needed by the receiving society
- safe third-country policies (e.g. as used by Germany in collaboration with Bulgaria, Poland, Romania and the Czech Republic to prevent land movement from these countries without a valid visa)
- temporary protection followed by repatriation once crisis subsides (e.g. used for many Bosnians in western Europe in 1996–97)
- policies of NGOs, some governments (e.g. Sweden) and countries of origin to address the deep-seated causes of poverty

Stateless peoples and ethnic cleansing

The Universal Declaration on Human Rights states that everyone has the right to a nationality. However, there are peoples for whom this right is disputed and who are denied security to find work and a home. The largest group in Europe is the Roma gypsy population; there are, for example, 1.8 million in Romania and 700 000 in Bulgaria.

The break-up of Yugoslavia and the Soviet Union also resulted in some groups being declared stateless for political, cultural and nationalistic reasons (e.g. Russians in Latvia and Serbs in Croatia and Macedonia). The result has been **ethnic cleansing** — the forced and voluntary movement of peoples to states that are culturally compatible (having the same religion and/or language). Where the patchwork of cultural groups has been too complex, there has been some internal ethnic cleansing to produce areas dominated by one culture.

Other cases of statelessness that could be studied include the 3 million Palestinians in Gaza and the West Bank, the Vietnamese in Cambodia, the Senegalese in Mauritania and Myanmari Muslims.

Migrant labour

In 1991, Jersey had 40 751 persons born outside of the island, of whom 76% were born elsewhere in the UK, 8% in Portugal, 5% in Ireland, 3% in France and 8% in other countries. The population pyramids for Portuguese and Irish show several characteristics of labour migrants. Those aged 20–35 dominate at 53.4% of the migrant labour population (28.5% are male and 24.9% female). There are more male migrants. In contrast, females dominate among the 20–34-year-old Irish migrants (25.7%) whereas similarly aged males form only 21.1% of the Irish.

These differences probably reflect a number of factors:

- The Irish have a language affinity with the islanders and were early migrants.
- Distance from the home country has meant that males are more likely to be the dominant Portuguese migrants. In addition, many Portuguese migrants to the Channel Islands come from islands, such as Madeira, that are even further away.
- It is a mainly single population judging from the small dependent population of children and retired persons.
- Both countries have higher unemployment and lower living standards than Jersey.

Table 4 Permanence of migration to Jersey

Year of arrival	Irish-born (% of 1991 total)	Portuguese-born (% of 1991 total)	UK-born (% of 1991 total)
1985	2.4	2.5	2.5
1986	4.2	2.3	2.8
1987	6.6	3.5	3.5
1988	6.7	2.9	4.0
1989	9.1	3.6	4.8
1990	13.4	13.4	7.0
1991*	4.6	24.3	0.2
Percentage of these arrivals still present at time of census	47.0	52.5	24.8

*Percentage of 1991 total arriving in first four months before census.

The Portuguese are more recent and short-term migrants, often returning home at the end of the tourist season. The same characteristic is repeated, but less strongly, for the Irish, who have a longer association with the island. Both groups are employed in lower-skilled work, particularly in hotels, although some of those who have settled in or keep returning to the island have risen to the status of managers in the holiday industry. UK-born immigrants are a mixture of affluent people seeking a tax haven and skilled workers attracted to work in the finance sector and in government employment such as teaching. Therefore, the UK-born are less likely to be recent migrants.

One of the consequences of the Portuguese presence is that many of the shops in St Helier (e.g. Boots) have Portuguese-speaking assistants to serve the immigrants.

Other cases where guest workers make an important contribution to the national economy are: Switzerland (19% of the population and 17.3% of the labour force); Germany (8.9% of the population and 7.1% of the workforce); France, where many have migrated from the former colonies in North Africa; and the Gulf States such as the United Arab Emirates, Bahrain and Kuwait, where mainly males have migrated from India, Pakistan and Bangladesh. Similar movements can be found in southeast Asia: from Indonesia to Malaysia and from the Philippines to Singapore and Malaysia.

Table 5 shows some of the costs and benefits of migrant labour to both the source and destination countries.

Table 5 Issues resulting from labour migration

Source country		Destination country	
Economic costs	**Economic benefits**	**Economic costs**	**Economic benefits**
Loss of young adult labour to source country	Less underemployment	Cost of educating children	Many less desirable jobs taken in host country
Loss of skilled labour may slow development	Return migrants (reverse or counter-migration) bring skills back home	Some skills may be bought cheaper, so home people are disadvantaged	The host country may gain skilled labour at little cost; can fill skills gap (e.g. programmers from India to USA)
Regions of strongest out-movement suffer from spiral of decline (e.g. parts of inland southern Italy)	Funds repatriated to source country	Funds repatriated to source country; pension outflows following retirement to home country	Some retirement costs transferred back to source country
Loss of skilled labour may deter inward investment	Less pressure on resources such as food	Slightly greater pressure on resources	Dependence of some industries on guest workers (e.g. construction)
Social costs	**Social benefits**	**Social costs**	**Social benefits**
Encourages more to migrate, with effect on social structure (e.g. Ireland in the 1840s)	Reduced population density; lowers birth rate because young tend to migrate	Discrimination against ethnic minorities (e.g. riots in Bradford, 2001); political unrest and opposition from parties of the right (e.g. in Austria)	Creation of multi-ethnic, multicultural society — building cultural facilities
Dominance of females left behind	Funds sent home can provide for improved education and welfare	Dominance of males, especially in states where status of women is low (e.g. Gulf States)	Greater awareness of other cultures (e.g. Caribbean music, literature and religion in the UK)
Non-return can unbalance population pyramid	Retirees building homes	Loss of aspects of cultural identity, especially among second generation	Providers of local services (e.g. Turkish baths and UK news-agents)
Returning on retirement is a potential cost	Some able to help develop new activities (e.g. tourism)	Creation of ghettos (e.g. Turks in Berlin); schools dominated by immigrants	Growth of ethnic retailing and restaurants (e.g. Rusholme, Manchester)

Introducing the global economy

Economic groupings

Economic characteristics of countries are not easy to pigeonhole. Table 6 shows that countries in a particular grouping are not always similarly ranked according to every criterion.

Table 6 Characteristics of the major types of national economy

	G7	MEDC	NIC	Oil-rich	Former Soviet	LEDC	Least developed
Exemplar countries (HDI rank 2000)	USA (6) Japan (9) Germany (17)	Spain (21) Greece (23) Belgium (5)	Malaysia (56) Thailand (66) Mexico (51)	Bahrain (43) UAE (40) Kuwait (43)	Hungary (36) Poland (38) Bulgaria (57)	Bangladesh (132) Ghana (119) Bolivia (104)	Rwanda (152) Burkina Faso (159) Tanzania (140)
GDP per capita ($), 1999	31 872 24 898 23 742	18 079 15 414 25 443	8209 6132 8297	13 688 18 162 21 875	11 430 8450 5071	1483 1881 2355	885 628 501
Agriculture % of GDP (% of labour force)	n.a. (3) 2 (7) 1 (4)	4 (12) 7 (23) 1 (3)	11 (27) 10 (64) 5 (28)	n.a. (20) 2 (8) 1 (1)	6 (15) 3 (27) 15 (13)	25 (65) 36 (59) 18 (47)	46 (92) 31 (92) 45 (n.a.)
Industry % of GDP (% of labour force)	n.a. (26) 36 (34) 28 (38)	28 (33) 20 (27) 25 (38)	46 (23) 40 (14) 28 (44)	n.a. (30) 64 (27) 52 (28)	34 (38) 31 (36) 23 (48)	24 (16) 25 (13) 18 (18)	20 (3) 28 (2) 15 (n.a.)
Services % of GDP (% of labour force)	n.a. (71) 62 (59) 64 (58)	69 (55) 72 (50) 73 (70)	43 (50) 50 (22) 67 (28)	n.a. (68) 35 (65) 47 (74)	61 (47) 65 (37) 62 (38)	50 (18) 39 (28) 64 (36)	34 (5) 40 (6) 40 (n.a.)
Urbanisation (2000 % of population)	78 78 88	78 68 97	58 22 78	92 86 98	67 67 73	21 39 65	7 38 28
UNDP rank in gender-related development index (GDI), 1999	4 11 15	21 24 7	55 58 49	41 45 40	35 36 53	121 108 94	135 144 124

Note: all rankings are out of 162 countries in the UNDP *Human Development Report*, 2001.
Data for the three countries in each class are repeated in the same order in the boxes below.

Some authors add **recently industrialising countries** such as Chile. The data suggest that, at best, there is a **continuum** of status rather than easily identifiable categories.

Figure 18 shows one way to map the economies of the world. Even here, not all have been mapped for lack of data.

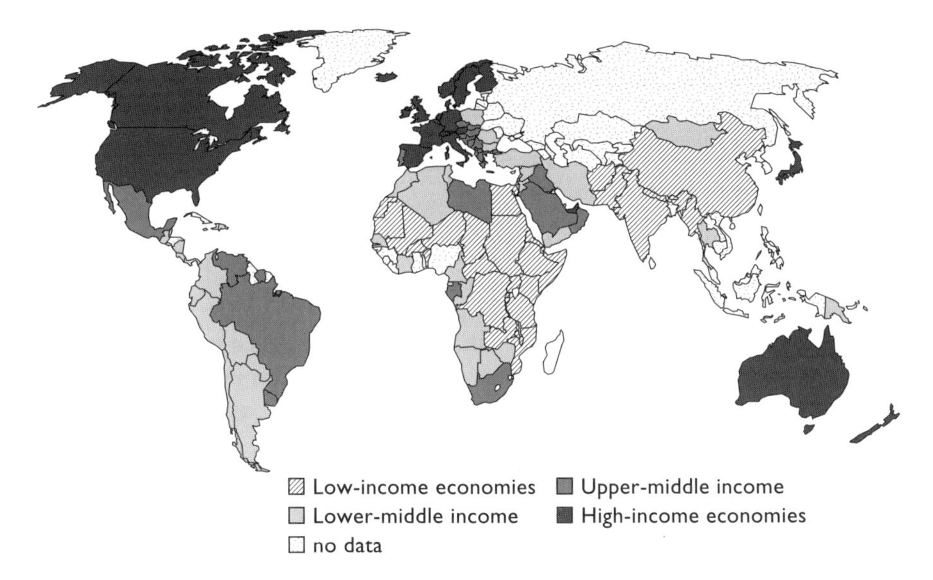

Figure 18 The four groups of world economies

Other authors talk of the **First World** (developed), **Second World** (former communist) and **Third World** (developing). The term **north–south divide** was coined by Willy Brandt for the Independent Commission on International Development in 1977. The **development gap** — the increasing disparity in levels of development — is the product of the uneven pace of development between the north and south.

Many countries from the groups above have worked together in various ways to protect and enhance their economic status, domestic economy and level of development:

- **G7/8** (1975) is a group of the wealthiest countries which now includes Russia.
- **G10** or **Paris Club** comprises the wealthiest members of the International Monetary Fund.
- **G15** (1990) is made up of the nations which meet to discuss Third World issues.
- **G77** (1975) (originally 77 and now 128 countries) pressurises the developed world for aid.
- **OPEC (Organisation of Petroleum Exporting Countries)** (1960) is a Vienna-based **cartel** representing interests of oil exporters, but it has grown to protect the interests of the Third World. Its strength is undermined by countries such as the UK and Norway, which are not members and do not agree with OPEC-controlled prices.
- The **European Union** (formerly the European Community) (1965) is the name adopted in 1992 for 15 states in Europe that have signed up to the 1957 Treaty of Rome to establish a single market for goods and services, and a coordinated set of policies for economics, defence and home affairs (such as immigration).

- **NAFTA (North American Free Trade Agreement)** (1994) is the largest established market for goods and services, and includes the USA, Canada and Mexico.
- **ASEAN (Association of South East Asian Nations)** (1967) is a regional alliance, based in Jakarta, designed to foster economic growth in, for example, Indonesia, Thailand, Malaysia, Myanmar and Singapore.
- **LAFTA (Latin American Free Trade Association)** (1960) was founded and relaunched in 1980 to remove all trade restrictions among members.

The last four are all **trade blocs**.

A set of key organisations that have been established to regulate economic development are listed below:

- **GATT (General Agreement on Tariffs and Trade)** (1948) was set up by developed countries to protect their trade advantages. It was replaced in 1995 by the **WTO (World Trade Organisation)**. This gave GATT a broader remit to deal with trade in services and intellectual property as well as goods. It is often seen as being more in the control of TNCs and promoting globalisation.
- **OECD (Organisation for Economic Cooperation and Development)** (1961) is a group of 25 nations aiming to raise living standards and encourage economic expansion in members and non-members.
- **IBRD (International Bank for Reconstruction and Development)**, also called the **World Bank** (1944), was set up as an aid to post-1945 reconstruction. It aids development but is often seen as a tool of the developed countries and is regarded by the south as **neo-colonial**.
- **IMF (International Monetary Fund)** is linked to the IBRD and GATT as a means to assist the world economy. Many see it as an instrument of US domination because, until 1994, it issued its funds in dollars.

Global interdependence and globalisation

Global interdependence is the way in which economies are all interlinked. A problem in one country can have repercussions on others, as was illustrated by the impact of the terrorist attacks of 11 September 2001, which affected airlines, aircraft manufacturers and airport workers across the world. Much of our lifestyle today is dependent on primary products from across the world; exotic vegetables from Thailand is but one example. Another example is plasma-screen technology which was developed in the USA and Japan, initially manufactured in those countries, but subsequently manufactured in South Korea. Clothing styles are designed in Milan, exhibited in New York and copied to be mass-produced in Bangladesh.

Globalisation is defined as close economic interdependence between the leading nations in trade, investment, and cooperative commercial relationships, and in which there are relatively few artificial restrictions on the cross-border movement of people, assets, goods or services. It is a stage in the economic development of the world caused by the development of global corporations (**transnational corporations** (TNCs) and **multinational corporations** (MNCs)) with more economic power than many states. It is not a new phenomenon. Companies such as the East India Company and the Hudson Bay Company in eighteenth-century London operated in a global market. Modern globalisation grew after 1975 as money from OPEC was invested in the developed world's banks, which then funded TNCs to develop the resources of the Third World.

There are three forms of globalisation:

- **economic** — the growth of TNCs at the expense of national governments
- **cultural** — the growth of western influence over aspects of culture such as music, the arts and the media
- **political** — the growth of western democracies and their influence, including the decline of centralised economies after 1989

Globalisation is promoted by:

- ICT, enabling rapid movement of money, ideas and information
- cheaper transport costs
- free market economics linked to the spread of democratic ideals
- the role of the **World Trade Organisation** and **trading blocs** promoting **free trade**
- the rise of **multinational corporations**, which are responsible for 80% of foreign investment and 50% of world trade
- new products (e.g. the first CD, 1979) and production methods
- the increased importance of **services** such as insurance
- the impact of **aid**

Impact of globalisation

An important effect of globalisation has been to widen the **development gap**, so much so that a 1999 report attributed the debt crisis to its impact on the economies of the poorest states. MEDCs, which have become richer as a result, like to call the new order the **global village**.

A second consequence has been **global shift** — the movement of production and some services to low-cost locations away from MEDCs. It results in deindustrialisation in MEDCs and industrialisation in NICs, RICs and LEDCs. In addition, patterns of inward investment still favour MEDCs because the MEDCs are still the source of innovation and the location of highly qualified labour.

Finally, globalisation has an impact on the environment, linking to the themes discussed in the first part of this unit, for example global warming, biodiversity, biodegradation, environmental pollution and environmental sustainability.

Trade flows in the global economy

Bananas

Bananas are a nutritious food that grows rapidly. Around 25% of production is traded and the rest is eaten in the areas of production. The banana trade illustrates many of the issues of globalisation (Table 7, overleaf).

Table 7 Banana wars in the 1990s

USA vs EU		
The trigger		• Protecting trade with former colonies — the Lome Convention • Tariffs on Central American bananas
The armies	• American finance group owns **Chiquita** • Texas Pacific group owns **Del Monte** • Third producer is **Dole** • These MNCs control 80% of production	• Main trading companies are **Geest** and **Fyffes** • These buy from farmers in the Caribbean
The territory	• Costa Rica, Guatemala, Colombia and Ecuador have extensive, **commercial plantations**	• St Lucia, Dominica, Dominican Republic, St Vincent, Grenada, Martinique (Fr) — all small • 10 000 farmers depend on crop in Dominica, which is 80% of the GDP
The costs of production	• Low costs — $3 per box • Wages $28 per week	• High costs — $11 per box • Losing markets
The logistics	• MNC's own freighters, distributing and processing companies	• Sell to the companies that have their own freighters etc.
Backing for war	• US government went to World Trade Organisation (WTO) • Democrat Party was partly funded by the banana producers	• EU trying to help its former colonies to maintain trade links • Companies and small producers have political clout
Natural disaster impacts	• Hurricanes forced farmers to reduce labour costs because production was affected • Bananas became cheaper	• Hurricanes destroyed the smallest farmers who had no money to reinvest • Increased costs for those who could reinvest
WTO decision	• Supported the US case	• EU tariffs on Central American bananas declared illegal
Responses	• Various environmental and community programmes were started by companies to improve their image • Company websites have all the details	• Organic farming of bananas commenced in 1994, often on former banana land (once clean of fertilisers) • Organic sprays (e.g. citrus acid) used • 'Fair Trade' bananas, but problem over import licences for small quantities • An example of **sustainable development**

Clothing

The Timberland Company was founded in New Hampshire, USA, in 1955 to produce leather boots and has diversified into outdoor clothing. Today, the company has 5550 staff working in retailing across the globe and $1.1 billion of revenue. It sources its products mainly from LEDCs, NICs and RICs for sale in 30 countries. Distribution is through depots in Kentucky and Enschede, Netherlands, to stores and factory outlets mainly in North America and Europe. The company now retails in Asia. It aims to build the company into 'an integrated global lifestyle brand'.

'Timberland Man' (Figure 19) illustrates the wide range of sources. Goods are manufactured where labour is cheap, and retailed in countries where the prices of the goods and the profit margins are high. The ultimate beneficiaries are not only the controlling Swartz family, which owns much of the company, but also the shareholders.

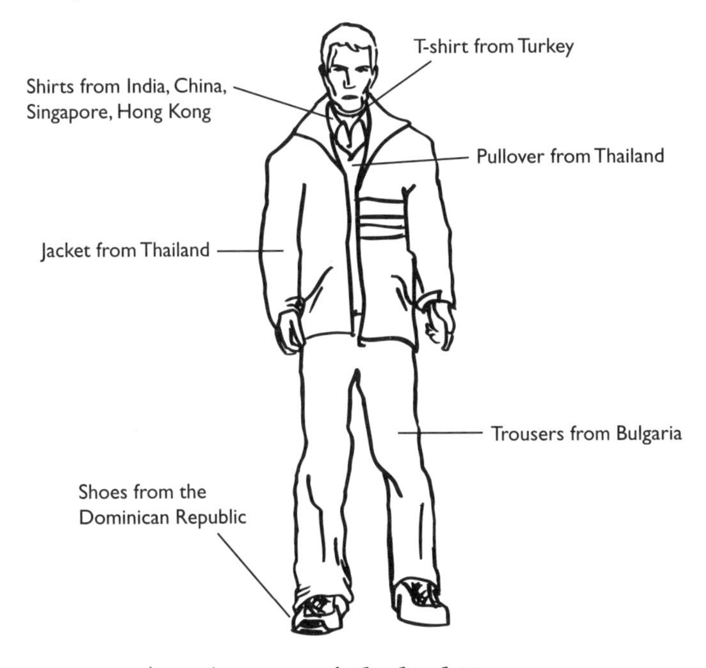

Figure 19 Timberland Man

Trade in the world is predominantly between MEDCs (Figure 20). However, the essence of the development gap is that MEDCs specialise in those products where they have the greatest advantage (**comparative advantage**). In addition, there is an **international division of labour** arising from comparative advantage. For example, Timberland's management is in the USA, whereas its production takes place where there is cheap labour and/or cheap raw materials. The resources of the south — labour and primary products such as cocoa and copper — are being sent to the MEDCs, where they have high value added. Some products are sold back to the LEDCs at great cost to the people who produce the raw materials.

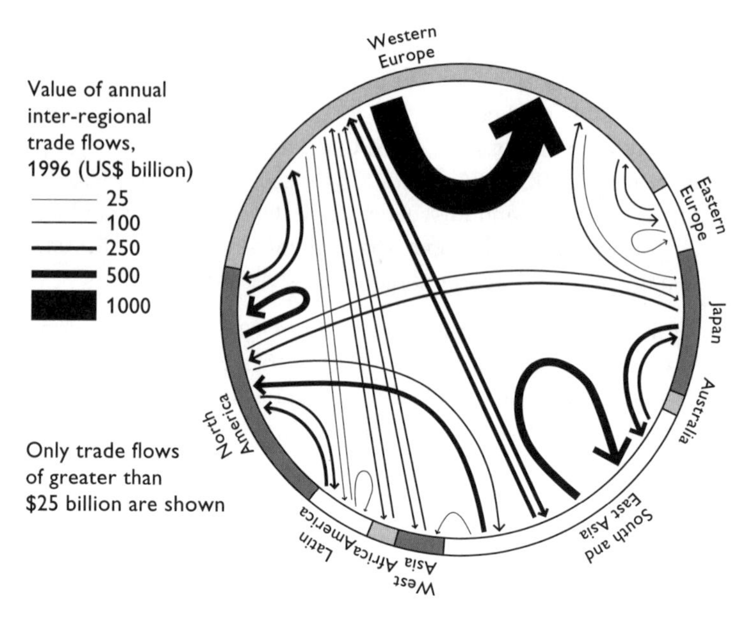

Figure 20 The value of annual inter-regional trade flows, 1996 (US$ billion)

Foreign direct investment

Foreign direct investment by major corporations is seen as a way of reducing the development gap. Figure 21 shows the level of investment of a TNC. Other forms of investment might be single projects related to aid, for example major dams.

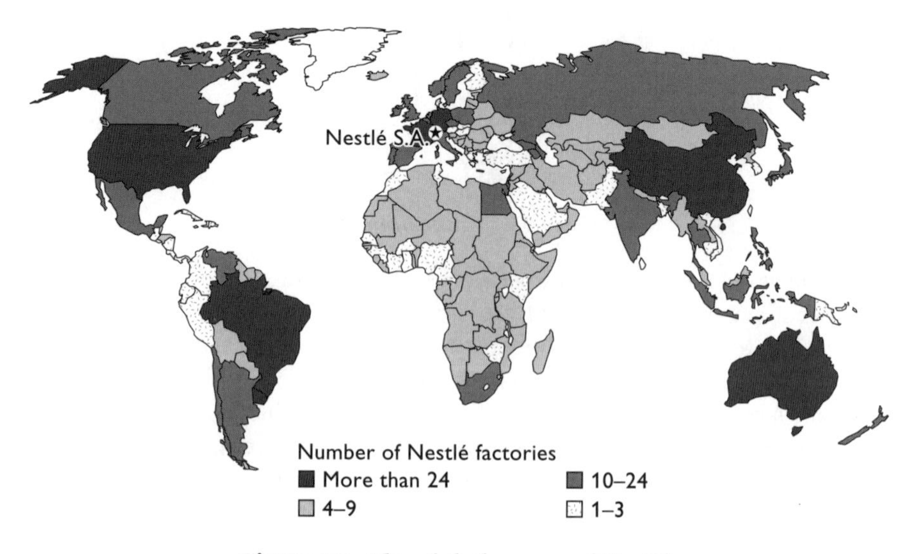

Figure 21 The global power of Nestlé

Aid

Aid, or **official development assistance (OCA)**, is another method of investment in developing economies.

Types of aid

The types of aid are not mutually exclusive, but can be categorised as follows:

- **Bilateral aid** is from country to country in the form of grants, accounting for 75% of all official aid. Half of this aid finances projects of interest to the donors.
- **Multilateral aid** is channelled through agencies such as the World Bank and IMF.
- **Food aid** accounts for 10% of all bilateral aid.
- **Disaster relief** accounts for 2% of all bilateral aid, and is not targeted on projects.
- **Aid and trade provision** is a form of bilateral aid used to prioritise industrial development projects.
- **Debt relief** is a form of **non-project aid**.
- **Programme aid** supports vital imports when countries cannot pay for goods.
- **Project aid** is tied to a specific project.
- **Tied aid** limits the receiver to buying from the donor country, whereas **untied aid** enables procurement from anywhere.

Who benefits from aid?

Aid is a process in which 'you collect money from poor people in the rich countries and give it to the rich people in poor countries' (Schumacher, 1974). Aid given in the past 25 years is less than the interest and loan repayments paid on the money. Much aid is for big projects such as dams, where the beneficiaries are the TNCs and MNCs that build dams and use the power. People in the LEDCs are moved away from fertile areas being flooded, while downstream water flow is reduced.

Globalisation and changing economic activity

Global shift and the relocation of industry

Industry between the 1920s and 1980s was **Fordist**. It relied on production lines in major industrial cores, for example the northeastern USA, the Ruhr in Germany, and the West Midlands. Major mass-producers established **branch plants** as they expanded, capturing larger markets at home and overseas. If the market grew large enough, a separate division was established, for example Ford, Europe. The result was a complex web of cross-border flows of parts to key assembly lines. Labour had been divided spatially and internationally.

Between 1951 and 1961, the West Midlands had rising employment (14%; UK, 8.5%) driven by growth in car production and also metal goods, and mechanical and electrical engineering, which were invariably linked to car production. However, between 1970 and 1990, 50% of output and 0.5 million jobs were lost as a result of overseas competition and cheap imports — a process known as **deindustrialisation**. This left a legacy of smaller firms, industrial dereliction, out-of-date buildings and unemployment. People had limited skills because they worked in routine, monotonous jobs on production lines.

New industrial systems are different and are summarised in Table 8.

Table 8 Changing production systems

Characteristic	Fordist industry, 1920–80	Just-in-time, post-1980	Flexible
Technology	Standardised parts; costly to retool for new products	More flexible and capable of rapid change	Simple, flexible machinery; not standardised
Labour	Narrow skills base; repetitive work; semi-skilled	Multi-skilled, flexible workforce; team-working and task-switching; more robotic production; feminisation	Highly skilled craft-workers dependent on modern technologies to support production; feminisation
Linkages	Independent suppliers, with large stores of parts in case of disruption	Just-in-time deliveries, with strong contracts	Close contact with suppliers and customers using ICT
Production	High volume of standardised products	High volume of distinctive products that cater for individual tastes	Low-volume, but highly customised according to demand
Examples	Vauxhall/Opel (GM) at Luton (closed 2002) and Russelsheim, with branch plants on Merseyside and Bochum	Toyota in 24 countries throughout the world (global shift from Japan); it is part of a big conglomerate or **keiretsu**, making everything from steel to textiles, and with a banking arm	Benetton makes high-value-added products near Venice which it distributes using ICT to ascertain market demand and respond rapidly; it sub-contracts labour-intensive work to 500 cheap labour locations in central Italy; now has its own bank and a Formula 1 racing team

Toyota is one part of a very large and complex organisation, with its headquarters in Tokyo but with interests in all fields of modern manufacturing (Figure 22).

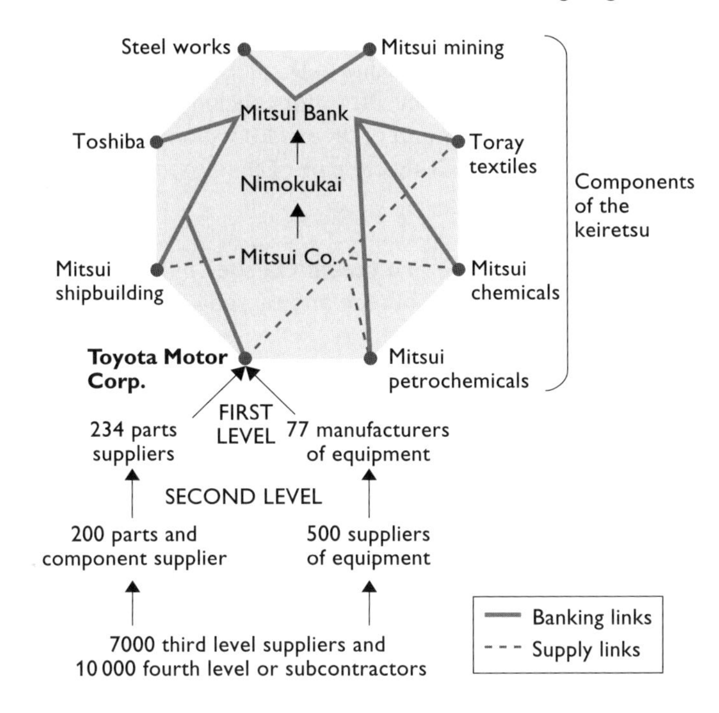

Figure 22 Toyota's position within the Nimokukai keiretsu

The case of Silicon Valley: a global innovation region

Silicon Valley, California, the popular name for the Santa Clara Valley, extends from San Francisco International Airport south to San José. The valley is home to the headquarters of Hewlett-Packard, Intel, Cisco, Apple, Yahoo and 3COM. In addition, there are many other biotechnology, software, computer and internet firms. The reasons for this concentration are as follows:

- Hewlett and Packard both studied at Stanford University, in Palo Alto, and began their business in a nearby garage.
- Jobs was hired by Hewlett before he founded Apple.
- **Cumulative causation** enabled the successful companies to spin off further successes.
- **Venture capitalists**, those who invest money in new start-ups, recognised the concentration of talent in the universities, such as the Santa Clara University Centre for Science, Technology and Society, and in the companies that followed Hewlett-Packard.
- High salaries and the lifestyle attracted others from the USA and beyond.
- In 2001, economic slowdown and crises in energy supplies slowed growth and gave rise to company failures. Plants located worldwide, but operated from Silicon Valley, have often borne the brunt of the slowdown.

The case of Silicon Glen: global shift between G7 countries

The 'silicon' tag has been applied to many areas of high-tech growth: **Silicon Fen** (Cambridge), **Silicon Thames** (the M4 corridor), **Silicon Alps** (the Munich region) and **Silicon Glen** (central Scotland from Erskine to Dundee — see Figure 23). The companies were attracted to Scotland by several factors, including:

- **regional development loans** as part of UK and EU regional policy to attract **inward investment**. The new companies were part of the programme to counteract the deindustrialisation of Scotland.
- relatively **cheap labour**
- availability of both **brownfield** and **greenfield** sites in the new and developing towns of the urban fringes of Edinburgh and Glasgow

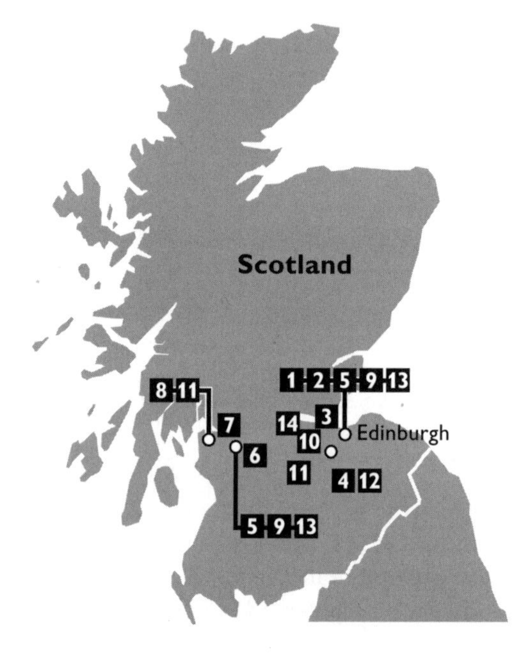

1. **3Com**, Edinburgh — 5200 job losses worldwide
2. **Adobe**, Edinburgh — 247 job losses worldwide
3. **Agilent**, South Queensferry — 8000 job losses worldwide
4. **Cadence**, Livingston
5. **Cap Gemini**, Edinburgh and Glasgow — 5400 job losses worldwide
6. **Cisco**, Bellshill — 5000 job losses worldwide
7. **Compaq**, Erskine — 700 Scottish job losses
8. **IBM**, Greenock — 1000 job losses worldwide
9. **Microsoft**, Edinburgh and Glasgow
10. **Motorola**, Bathgate — closed with the loss of 3100 jobs
11. **National Semiconductor**, Greenock — 1000 job losses worldwide
12. **Oracle**, Edinburgh and Glasgow
13. **Sun Microsystems**, Linlithgow — 4000 job losses worldwide

Figure 23 Job losses announced in Silicon Glen, 2000–01

All of the companies that came to Scotland were branches of major global companies, such as IBM (Greenock), Motorola (Bathgate), Adobe (Edinburgh) and Compaq (Livingston). The 2001 economic downturn threatened the Glen's branch plants. NEC at Livingston mothballed its semiconductor plant, which employed 1250 people. Motorola was closed with the loss of 3100 jobs. Compaq lost 700 jobs. In 2001, 12000 Scottish electronic industry employees lost their jobs.

TNCs: the major players in global shift

The UN defines TNCs as corporations that 'possess and control means of production or services outside the country in which they were established'. Their size is measured by revenues, market capitalisation and, sometimes, employees (see Table 9).

Table 9 The world's 25 largest companies, ranked by market capitalisation

Rank (2001)	Company	Country	Employees	Sector
1 (1)	General Electric	USA	313 000	Electronics, electrical and aircraft engines
2 (5)	Microsoft	USA	39 100	ICT
3 (3)	Exxon Mobil	USA	79 000	Oil/chemicals
4 (6)	Wal-Mart	USA	1 244 000	Retail
5 (7)	Citigroup	USA	242 000	Banking
6 (4)	Pfizer	USA	90 000	Pharmaceuticals and biotechnology
7 (9)	Intel	USA	86 100	ICT
8 (15)	BP	UK	98 000	Oil/chemicals
9 (24)	Johnson & Johnson	USA	n.a.	Pharmaceuticals
10 (10)	Royal Dutch/Shell	Netherlands/ UK	95 000	Oil/chemicals
11 (11)	American International	USA	55 000	Insurance
12 (18)	IBM	USA	316 303	ICT
13 (19)	Glaxo-Smithkline	UK	108 201	Pharmaceuticals
14 (16)	NTT DoMoCo	Japan	15 100	Telecommunications
15 (13)	Merck	USA	n.a.	Pharmaceuticals
16 (22)	Coca-Cola	USA	36 900	Beverages
17 (8)	Vodafone Group	UK	29 465	Telecommunications
18 (17)	SBC Communications	USA	220 090	Telecommunications
19 (21)	Verizon Communications	USA	n.a.	Telecommunications
20 (2)	Cisco Systems	USA	34 000	ICT
21 (41)	Procter & Gamble	USA	n.a.	Personal care and household
22 (26)	Novartis	Switzerland	n.a.	Chemicals
23 (28)	Home Depot	USA	140 000	Retail
24 (46)	Philip Morris	USA	n.a.	Tobacco
25 (33)	Total Fina Elf	France	210 709	Oil/chemicals

Source: *Financial Times*, 10 May 2002.

The data in Table 9 are only a snapshot in time, but they do illustrate the following points:

- Most top companies are American.
- Both manufacturing and service industries are present.
- Not all are big employers — the service sector companies have the highest number of employees.
- No automobile companies appear in this list (Toyota was 25th in 2001). All of the companies in this list could be studied as examples of TNCs, to establish their pattern of growth and production, and their impact on the environments in their country of origin and where they have branches.

Significantly, the headquarters of TNCs are concentrated in the 'North', as Table 10 illustrates.

Table 10 Headquarters of the top 500 companies by country and continent, 2002

Country	No. (%)	Country	No.	Country	No.
USA	238 (48%)	Spain	6	Denmark	2
Japan	50 (10%)	South Korea	6	Saudi Arabia	1
UK	41.5 (8.3%)	Mexico	6	UAE	1
France	28 (6%)	Sweden	5	India	1
Germany	21 (4%)	Singapore	5	**Continent**	**No.**
Canada	18 (3.6%)	Belgium	5	**North America**	**256**
Netherlands	14.5 (2.9%)	Russia	4	**Europe**	**155**
Switzerland	12 (2.4%)	Ireland	2	**Asia Pacific**	**72**
Italy	11 (2.2%)	Taiwan	2	**Latin America**	**8**
Australia	9 (1.8%)	Brazil	2	**Middle East**	**2**
Hong Kong	8 (1.6%)	Norway	2	**Australasia**	**9**

Toyota: a global automobile manufacturer

Toyota was founded in 1930 from a family weaving loom business, very similar to the origins of Morris in the UK and Opel in Germany. From its headquarters in Tokyo, it manages a global manufacturing company employing 215 000 people in 41 subsidiaries in 24 countries. Once a Japanese-owned firm, it now has 14% foreign shareholders. The main manufacturing site is in the Takaoka area, where there are ten plants producing 700 000 vehicles a year using **just-in-time** technologies. It has a 25% share in its component makers, including Denso, the fourth largest parts maker in the world, which employs 85 000 people.

Stages in the company's growth

- 1950–75: developed as an **export-based** company from eight plants established in Japan

- 1959: JIT system installed
- 1959: Brazilian assembly plant
- 1967: tie-up with Daihatsu
- 1974: buying parts overseas began
- 1977: assembly in Ecuador
- 1986–98: move into US market **offshore production** — California (1986), Kentucky (1988, 1994), Arizona testing ground (1993) and Indiana (1998)
- 1987: technical centre in Belgium
- 1987: joint venture with VW in Hanover
- 1989–92: move into European market (Burnaston, Derby) to gain a foothold in the EU — 80% of 200 000 cars exported to Europe
- 1998: second European plant at Valenciennes — good communications in Europe and in euro zone
- 2001: alliance with PSA Peugeot Citroën for small car production — examining sites in Poland
- 2002: motorsports — Formula 1 team based in Cologne, Germany (and not in UK, like most other F1 operations)

Why is Toyota so successful?

- **R&D** carefully targeted to market needs and developed in each market
- **Just-in-time** production systems
- **Vertical integration** of suppliers closely tied to the company
- Strong technological support through joint ventures with Japanese computer manufacturers
- Automated, large assembly plants
- Labour management policies
- Japanese government support and both direct and indirect host government support when plants are established overseas
- Not much takeover activity among other manufacturers (VW owns Audi, Bentley, Bugatti, SEAT and Skoda), but alliances with specific targets, for example small cars for eastern Europe
- Assembly kits used where markets are smaller, for example Latin America

Some of these reasons may apply to other motor manufacturers.

Barclays Capital: a global service sector company

Barclays Capital is a prime example of a company operating in the global service sector. It functions as the investment-banking arm of Barclays Bank, and employs 3000 people at its headquarters in Canary Wharf, London.

The firm deals with £364 billion of investments through its 33 offices worldwide, managed by regional headquarters located in Paris, Frankfurt, New York, Tokyo and Hong Kong. Figure 24 (overleaf) shows the hierarchy of Barclays Capital offices worldwide.

Figure 24 Barclays Capital offices

The role of governments

Governments can influence the locations and types of economic development in various ways (Table 11).

In the UK, government can intervene in industrial location at three levels: European, national and local. From the 1930s to the 1980s, there was a raft of measures based upon spatially defined **Assisted Areas**, the issuing of **Development Certificates** that were strictly controlled in areas such as the southeast and **Regional Development Grants**. After the 1980s, the extent of Assisted Areas was restricted and grants were more selective. Today, development assistance is concentrated into the **Single Regeneration Budget (SRB)**. Almost 60% of assistance has gone to Wales and Scotland and 27% to the northwest and northeast. It has been used to attract **foreign direct investment (FDI)**. The EU **Structural Funds** are more focused on specific issues. For instance, **Konver** grants have been given for the regeneration of former military land released as a part of the peace dividend, for example Chatham dockyard and the area around Portsmouth harbour.

Table 11 The roles of national governments in economic development

	Market	**Social market**	**Planned market**	**Socialist planned**
Examples	USA, Japan	Sweden, Germany	Singapore, Malaysia	Cuba, China
Instruments	• Companies • The market dictates • Laissez faire	• Companies • Taxation • Government intervention	• State plans • Companies • Regulation • Single-party control	• State plans • Single-party control • Government decides • Strong role for ideology
Government provides	• Infrastructure • Basic education • Defence	• Infrastructure • Education • Health care • Defence • Planning framework	• Infrastructure • Education • Health care • Defence • Planning	• Total welfare system • Infrastructure • Defence • Planning
Public sector	Small safety net	Significant	Significant	Very large
Attitude to environment	Market will determine	Support for environmental legislation, e.g. Kyoto	Environmental legislation variable	Often neglected in drive to develop
Regional development	Left to market	Assistance available; pump priming of new activities	Highly planned	Highly planned

Foreign direct investment

Officially, FDI is where a company has at least a 10% interest in the investment in the receiving country. Figure 25 (overleaf) shows some key Japanese inward investment in the UK. Japanese inward investment is mainly in industrial sectors, such as cars and electronics. However, Japanese investment is not the largest. The USA and four European states invest more in the UK. FDI from these sources is frequently in the service sector, for example banking and insurance. The concentration of large, industrial FDI is in the strongest regions of the country. In 1999 London was the leading recipient of FDI in Europe with 136 schemes (Paris 74, Catalonia 73). Berkshire (M4 corridor) had 31 and was the second largest in the UK and ninth in Europe.

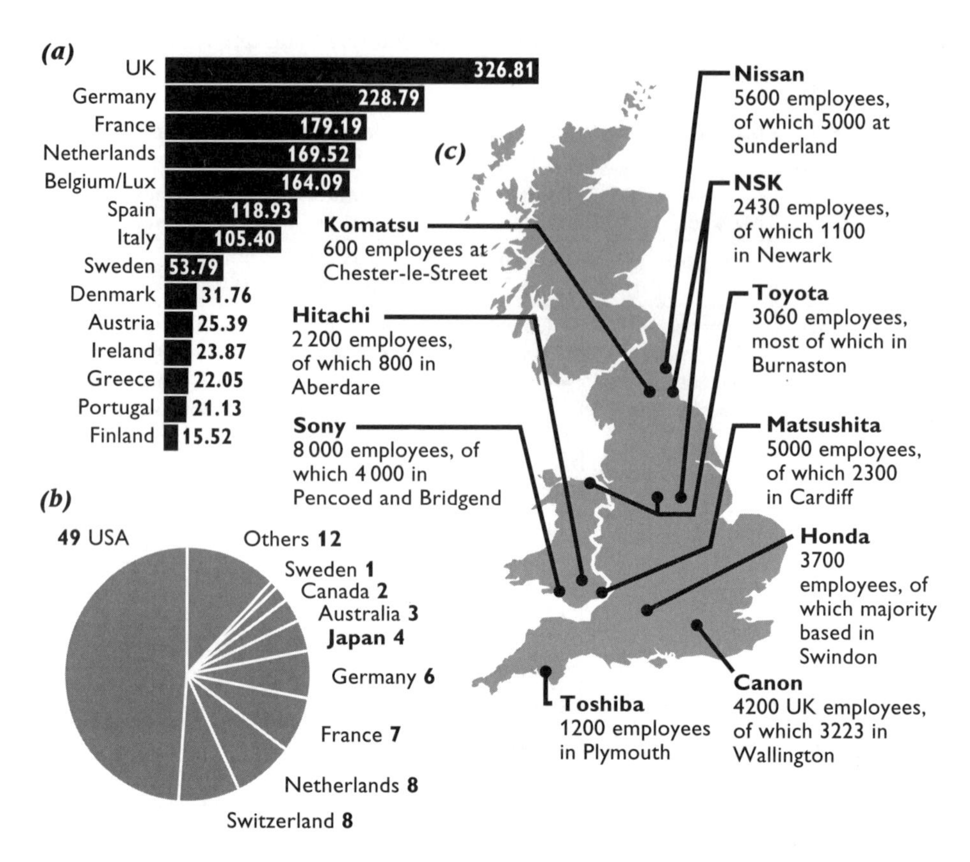

Figure 25 Japanese foreign direct investment: (a) share of investment in the EU ($ billion), (b) sources of foreign direct investment in the UK, (c) examples of Japanese inward investment

Why did the UK gain more FDI than elsewhere in Europe?

- The government created an economy with relatively less regulation than elsewhere in the EU.
- Regional assistance was available.
- English is an accepted business language, especially among the major investing MNCs.
- The UK's presence in the EU gave access to a major market. (The fact that the UK has not joined the euro zone is being used to explain a declining FDI rate since 2000.)

Benefits of FDI for the UK

- Regional regeneration — more than 50% of regeneration monies has been spent in the assisted regions.
- It creates exports. Of the top UK exporters, 37% are foreign-owned and these account for 40% of exports.

- New skills are gained through transfer from elsewhere.
- Some firms have focused activities in the UK to the detriment of EU neighbours, for example Hoover closed in Dijon to focus on Glasgow.
- For the firms, labour costs are low and regulations are fewer than elsewhere in the EU.

Deindustrialisation — a consequence of global shift

Deindustrialisation is the decline in manufacturing that has been experienced in the regions of North America and Europe that industrialised in the nineteenth and early twentieth centuries. This has come about for a number of reasons:

- Many products are at the end of their life cycle — newer products have replaced the old.
- Outmoded production methods have been replaced by newer technologies requiring less labour.
- Labour has clung to old practices of the Fordist economy. There has also been poor management of labour.
- There is competition from cheaper locations with low labour costs.
- There was a recession in the 1980s.
- Government support for industries such as coal and steel has been removed.
- Firms have needed to rationalise production.

British Steel (now Corus) closed the Ravenscraig steelworks in 1992 with the loss of 5000 jobs. The works were built to stimulate industrial growth in Scotland, but 96% of their products left Scotland. The prime reasons for closure were the need to rationalise production in South Wales and the northeast, the recession and the declining market for steel. In addition, the company was no longer state-controlled and needed to make a profit for investors. Industrial growth in the area was in high technology and not in heavy industry.

Newly industrialising countries

The case of Malaysia

Malaysia has gone through four phases of growth.

(1) Post-independence from the UK, Malaysia depended on colonial products — rubber, tin, palm oil and timber. Industries were geared to replacing imports, for example manufacturing rather than importing clothing.

(2) After 1970, oil exports provided tax revenues which were used by the Malaysian government to establish export industries. This was done by:

- breaking away from **neo-colonial** control over the economy, which was exercised by the firms extracting primary produce

- encouraging investment, for example by setting up an Investment Centre in Tokyo
- implementing state planning, which set targets for the country
- passing the Investment Incentives Act, encouraging FDI (60% of investment was foreign)
- creating **free trade zones** (e.g. Penang, 1972) to enable companies to develop without the burden of taxes
- creating **economic priority zones** to assist firms developing in the regions

It must be stressed that these developments were government- and not market-led.

(3) In the late 1980s, the government introduced a national plan to industrialise. Steel mills, cars (Proton), motor cycles, pulp and paper all developed on industrial estates near the federal state capitals (Shah Alam near Kuala Lumpur and Prai near Penang). Investment was encouraged, particularly from non-Chinese Asia (Japan, South Korea and Singapore) rather than from the West or China. This was a deliberate attempt to move away from the colonial past and to encourage the local population to dominate business life, rather than the Chinese, who had only migrated to the country in the nineteenth century. Tourism was developed, for example Penang Island. Investment in education was a final priority.

(4) Since 1990, the plans have focused on '20-20 Vision', the final thrust to achieve developed world status. However, recession at the end of the millennium slowed growth. Investments are in high-technology rather than electronics, for example Kulim Science Park (Intel) in Perak and the new cyber cities (Cyberjaya, with its IT university, and Putrajaya, the new governmental city linked to the new national airport at Sepang). There is a need to develop high-value-added products because labour is in short supply. Some routine production has been transferred elsewhere, for example Vietnam. There are plans to set up international growth triangles in collaboration with neighbouring states, to take advantage of FDI. Examples include Johor, where much investment comes from Singapore, and the Indonesia– Malaysia–Thailand triangle, where advantage can be taken of the less developed areas of Sumatra and southern Thailand for Malaysian investment.

Why has Malaysia been successful?

- State planning and *not* free enterprise was implemented. The government managed the economy and controlled FDI.
- The taxes available from oil in the South China Sea provided the initial funds for development.
- A 'visionary' leader, Dr Mahatir, led the government for two decades. His was a single-party government with strong state controls over the media.
- TNCs have clustered in FTZs and with other producers in the industrial estates.
- Labour and other laws favoured investors.
- Labour costs are lower than in MEDCs, but they are higher than in RICs, which could attract companies away in the future.

Problems resulting from NIC industrialisation

- Environmental degradation results from the exploitation of primary resources.
- There has been deforestation and conversion of vast areas for oil palm cultivation. A particular problem is 'the haze', a smog caused by burning forest as it is cleared, for example in Indonesia.
- Pollution occurs where controls are less stringent.
- There is greater divergence of earnings.
- More women are employed, which can reduce the birth rate to below the replacement level.
- Use of expatriate skilled labour, especially in the finance sector in Singapore and in unskilled sectors such as construction, means that distinct expatriate and immigrant areas develop.
- Migration of labour to rapidly growing areas is not just internal migration but also involves immigration.
- More training is available.
- The population has become westernised, though this has been resisted by the government in Malaysia.

Economic futures

Note: The three sections that follow are options. You need to study only one.

1 How might the global economy change in the future?

New markets

FDI has resulted in greater wealth creation in many NICs and RICs, whose populations have consequently demanded goods and services.

Beware of tables showing the fastest-emerging markets. The US government forecast that both Argentina and South Africa would be at the forefront of growth in 1997, yet by 2002 both were facing severe financial crises.

Emerging markets have large, relatively rapidly rising populations (e.g. Mexico, Indonesia, India and China) and have received large volumes of FDI. US investment has stimulated growth in Mexico where **maquiladora centres** — places that the Mexican government has permitted to import raw materials from the USA, free of duty, and re-export the finished product — have stimulated growth. In other cases, for example Poland, the strength of the former Soviet industrialisation has provided the opportunity for new FDI, often related to the car industry, to help raise incomes and demand.

New products

Figure 26 shows the **product life cycle**. Over the past 50 years, the time it takes for any product to be invented, developed, manufactured and superseded has reduced. The classic case in recent years has been the development of the mobile phone, from the cumbersome product of the 1980s to the miniaturised third-generation models. Compare the pace of development of the mobile phone with that of the gramophone/record player/music centre/CD player/MP3 player, which took over half a century. The time each new product dominates the market has dropped from decades to years. Some pharmaceutical companies that deal in the new biotechnologies have quickly become among the largest in the world.

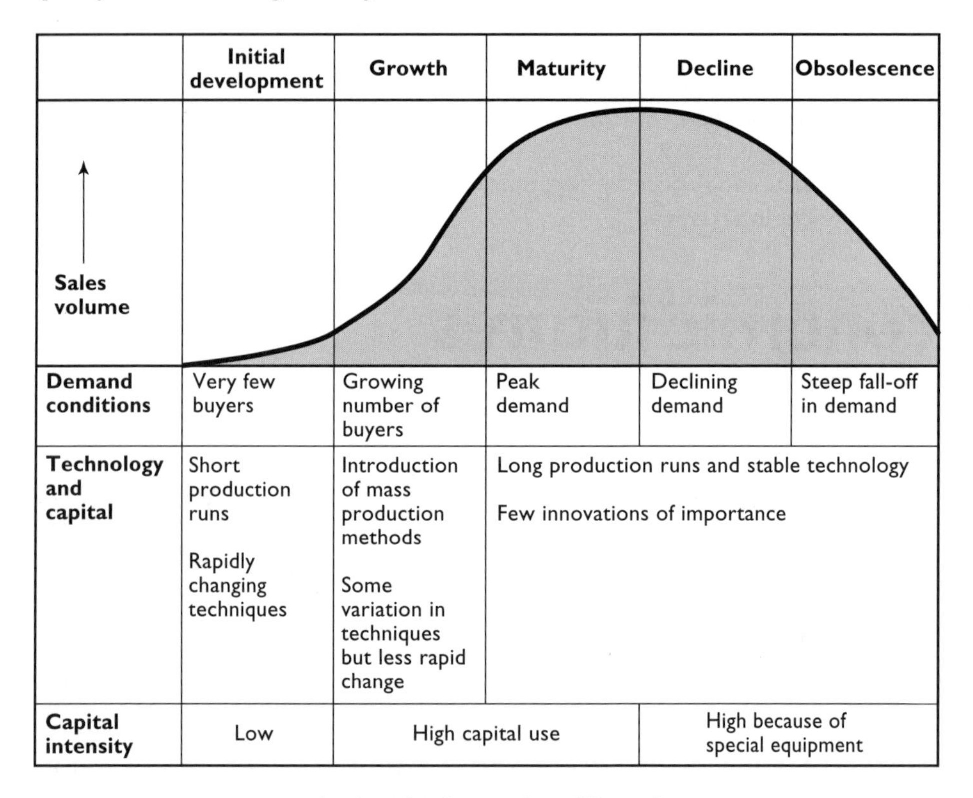

	Initial development	Growth	Maturity	Decline	Obsolescence
Sales volume					
Demand conditions	Very few buyers	Growing number of buyers	Peak demand	Declining demand	Steep fall-off in demand
Technology and capital	Short production runs Rapidly changing techniques	Introduction of mass production methods Some variation in techniques but less rapid change	Long production runs and stable technology Few innovations of importance		
Capital intensity	Low	High capital use		High because of special equipment	

Figure 26 The product life cycle

New production develops at key points in a series of **Kondratief waves** (Table 12). This option of the specification focuses on K5: ITC and biotechnology. While each wave has brought new countries into the more modern forms of production, the waves have generally resulted in new areas of production within countries.

As new areas develop, so many older industrial areas decline, e.g. the Rustbelt (the former northeastern industrial area of the USA), the Ruhrgebiet in Germany, the Nord coalfield in France and the British coalfield regions, such as west Cumbria and south Wales.

Table 12 Characteristics of Kondratief waves

Wave	K1	K2	K3	K4	K5
What (and when)	Early mechanisation (late eighteenth century)	Steam power (1840s)	Electrical engineering (1890s)	Fordist mass production (1930s)	ITC and biotechnology (1990s)
Infra-structure	Canals and turnpikes	Railways and shipping	Electricity grids	Roads and airlines	Digital networks and satellites
Where	Britain, France, Belgium	+ Germany, USA	+ Switzerland, Netherlands	+ EU, Japan, Sweden, Canada, Australia	+ Taiwan, South Korea, Singapore, China
Where in UK	Lancashire	Northeast	Arterial road estates, west London	Dagenham, Longbridge	M4 corridor, Cambridge

New services

Change here has been more rapid, with the rise of banking, insurance and investment as the **central business districts (CBDs)** of many cities have produced wealth and clustered their activities for ease of personal communication (e.g. the City of London and Manhattan). Pressure on space saw the decentralisation of offices in the 1960s and 1970s.

- Computerisation has reduced labour needs, but often further new activities, such as derivatives and futures trading, and foreign exchange dealing have employed the displaced (and more).
- More people own personal computers and have connected to the internet, with the consequence that e-based services, call centres and e-retailing are growing.
- There is more home-based working, involving two levels: low-skilled repetitive work and self-employed consultancy work (often among early retirees). Note the role of the **informal**, **black economy** (unregulated and outside the normal tax regime) and the **grey economy** (self-employed and variable tasks).
- There is more consumerism and more money to spend on, for example, clothing. Note the rise of designer brands such as Calvin Klein and Max Mara and the decline of mass-produced brands such as Marks and Spencer, and Burton. The rise of consumerism has an impact on cities.
- Over time, more money has been spent on travel and tourism. The 2002 growth areas are **ecotourism**, based on principles of **sustainability**, and cruising. Long-distance travel is increasing as costs fall and people search for new experiences.
- Budget airlines dependent on internet technologies and the standardisation of product and aircraft are growing service industries.
- The rise of TNCs and MNCs has resulted in more business travel and tourism.
- More women are working and returning to work after having children.
- New services have developed for workers, for example house sitting while awaiting deliveries.

- Entertainment away from home has increased. Restaurants, speciality bars and lap-dancing clubs all increase low-paid employment.
- Large stadia have developed for spectator sports and concerts.
- Fitness clubs have grown.
- Some services have transferred to LEDCs, for example the software industry for companies such as Hewlett-Packard, Motorola and Ericsson in Bangalore, India.

One theme that emerges is the division of labour between those who are at the cutting edge or in the economic cores on high salaries and those in peripheral, low-wage roles. These patterns occur at the global, national and regional levels.

The impact of global economic development

A number of challenges arise from the impact of development and change:

- **Non-renewable resources**, such as tin and copper, become exhausted.
- **Substitutes** are developed, for example fibre-optic cables for copper cables.
- Consumer product demand rises as the wealthy increase in numbers.
- People need the services to support new activities, for example banks, insurance companies and investment managers.
- **Pollution** increases, for example:
  - — CFC pollution as old fridges are discarded for newer versions
  - — higher demand for power, leading to greater carbon dioxide discharges
  - — disasters such as oil spills, chemical spills and slurry discharges
- There is increased energy consumption, despite energy efficiency drives, especially in industrialising LEDCs.
- People change careers/jobs more often and there is more flexible working, such as job sharing, flexitime and shift working.
- More people migrate in search of work.
- More people may not have jobs due to **automation** and **robotisation**.
- There is an increase in 24/7 economic activity, including teleconferencing to overcome time differences and travel costs.
- There has been a rise in the conference industry and in the demand for suitable venues.
- Protests against biotechnology, for example GM agriculture, are on the increase.
- The global triad (the USA, the EU and Japan) continues to dominate FDI, while the USA, the EU and southeast and east Asia still lead the way in manufacturing.

2 How might the development gap be addressed?

What causes the development gap?

- Primary products can be moved more cheaply. Shipping costs in 1990 were only 33% of those in 1920.

- Telecommunications costs in the same period fell by 90%. Internet subscribers are doubling every year.
- Trade barriers are fewer and their fall has stimulated trade.
- There are more capital flows around the world.
- More capital is controlled by MNCs.
- Technology is no longer associated with high productivity and high wages in MEDCs. Mexican productivity is rising due to **technology transfers** from the USA, but wages are not rising at the same rate. Technology is widening the development gap both between countries (Mexicans earn less to make the same product compared with those in the USA) and within countries (many less-skilled, routine tasks have been transferred overseas, while new technology and finance posts gain huge salaries).

Trade patterns, alliances and reforms

Figure 27 shows the structure of trade, which is dominated by the developed world (accounting for 75% of the world's exports and 81% of manufactured exports).

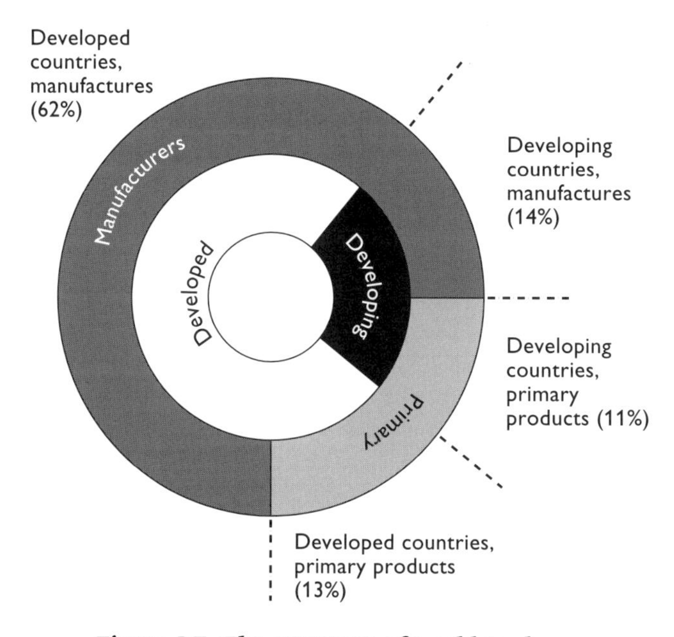

Figure 27 The structure of world trade

There are 32 trade blocs in the world. Most of the major ones, such as NAFTA and the EU, trade within themselves, whereas those in Africa, Latin America and Asia tend to trade more with the rest of the world. Besides trade, investment also flows around the world as FDI, portfolio investments in shares in companies and currency trading. The net gainers are often NICs, although the flow of profits is back to the MEDCs. Much trading in manufactured goods and services is controlled by TNCs, which have very intricate patterns of investment.

Reform is needed because countries, especially the most powerful, can act unilaterally if their interests are hurt. One example is the tariff placed on steel imports by the USA in 2002, in contravention of trade rules.

Weakness of regulatory bodies

The regulatory bodies of the new world order, which are dominated by capitalist states, are:

- international regulators (e.g. IMF and WTO)
- coordinating groups of countries (e.g. G7)
- regional blocs (e.g. the EU and ASEAN)
- national governments

Increasingly, however, the monetary and trading system is run by the institutions that manage the free flow of capital, i.e. the stock exchanges, national banks and money exchanges. It has been called **'casino capitalism'** because so much is based on activities such as **futures trading** (where parties to a transaction agree today the price at which a commodity will be traded in the future) and **speculating** on currency shifts. In theory, these activities are supervised by the Bank for International Settlements — a body developed by G7. But many think that all the bodies controlling trade are weak when faced by wealthy governments or big TNCs. Only when a dispute over terms of trade is between developed economies does a change of the rules come on to the agenda. As a result, 70% of the world's population are 'excluded' from the world economy, because they have assets that they are unable to realise to support development.

Reform initiatives

There have been two main types of reform initiative:

- **Top down** or **trickle down** is where responsibility for development is at a national level. This often involves large-scale investments resulting from aid. The Aswan Dam (Egypt), the Akosombo Dam (Ghana) and the Pergau Dam (Malaysia) are all examples of this approach. The approach was less successful than anticipated because the aid was often tied to the donor's economic and political interests.
- **Bottom up** is a more community-driven process in which the local population participates in deciding on the developments that best suit their needs. Many non-governmental organisations, such as Oxfam and Opportunity International (OI), work through local communities to assist development. OI is an international charity that loans money to enable people to start small businesses. For instance, a female chicken farmer and mother of four AIDS orphans in Kampala had only 150 chickens. With a loan of £108 she was able to expand her stock to 2000 birds and sell eggs in the market. She has repaid the loan, which is now available to enable another small-scale business to start. In Uganda, loans are available for AIDS orphans once they reach 14 years old. Other loans are available to employ 'AIDS apprentices'. The average loan from OI is £139 and 86% of loans are made to women. Over 95% of the loans are repaid and reinvested; as a result, 426 927 jobs have been created in Africa since 1981.

Reformers have proposed further changes to reduce the development gap:

- **Ex-President Bill Clinton** has proposed altering the WTO rules away from favouring the rich while aid to poor countries declines. Reform must encompass global environmental health and education policies.
- **George Soros** has suggested a transfer of wealth from rich to poor countries via a mechanism of donated special drawing rights, giving access to funds backed by a rich country.
- **Susan George** has proposed a move from Margaret Thatcher's TINA (there is no alternative) to TATA (there are thousands of alternatives), such as returning decision-making to local and national levels rather than leaving them with TNCs. She has argued for 'participatory budgeting' to give people what they want (e.g. in Porto Alegre, Brazil).
- **Hernando de Soto** has suggested giving people access to property rights over the land they occupy so that they can use this property as a means to obtain loans. This is being tried in Peru, Egypt and Mexico.
- **James Tobin** has suggested taxing currency transactions and using the money to assist development. This **Tobin tax** would help to iron out the worst fluctuations in the currency markets and threats to countries from currency speculators.

Some global social reforms that impact on the labour force are still only partially recognised. For instance, the UN Convention on the Elimination of Forced and Compulsory Labour (1957) is not recognised in Vietnam and Sri Lanka; the UN Worst Forms of Child Labour Convention (1999) is not recognised in Thailand; and the Minimum Age Convention (1973) has not been signed by Pakistan and Myanmar.

Aid versus loans

Official development assistance (ODA) comprises grants and loans. Most comes from OECD Development Assistance members to officially approved recipient countries. **Bilateral loans** require repayment in cash or goods and may include some grant element, whereas **bilateral grants** require no repayments. **Aid** is a broader term covering grants and loans. In the past, it was often **tied** to projects in which the investing country had an interest. Tying prevents misuse of arms, for example, but it often reduces the value of the aid. Today, a greater percentage (90%+) is **untied**.

Different donors have different patterns of aid (Figure 28, overleaf) related to colonial links (e.g. the UK), the region (e.g. Japan) and political support (e.g. the USA to Israel and Egypt). It has resulted in **aid dependency**, where aid forms a substantial proportion of national income. In 1997, ten countries took 75% of foreign investments; China alone accounted for 25%.

Table 13 (overleaf) shows how aid varies depending on which measure you take. Guinea-Bissau receives the highest as a percentage of national income, Mauritania the highest per capita, while China receives the largest total sum of the countries in the table.

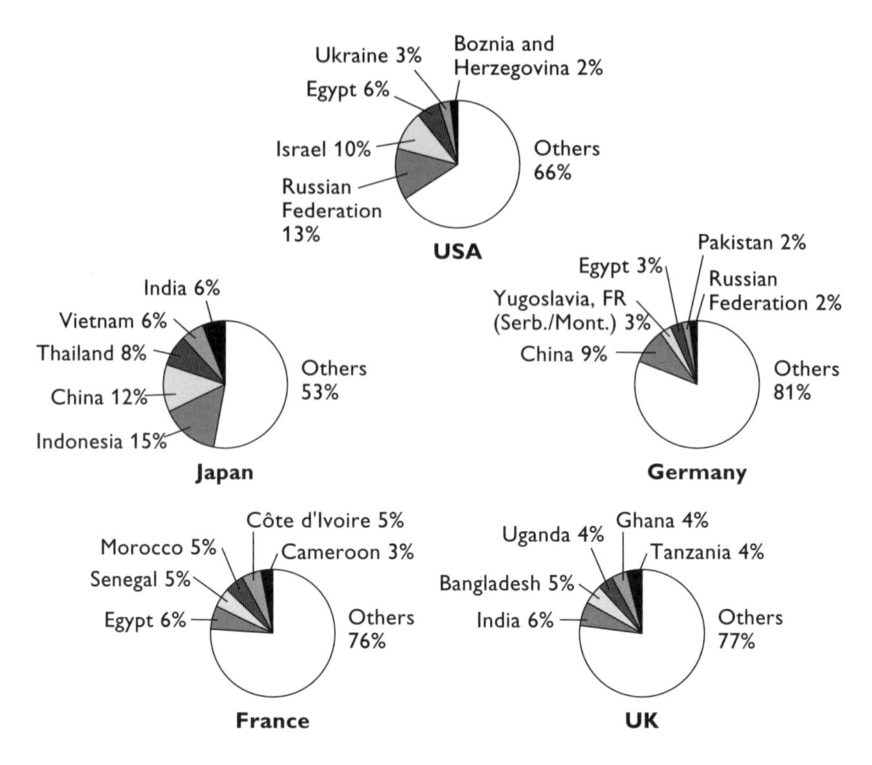

Figure 28 Destination of aid, 1999

Table 13 Some variations in aid receipts, 1994–99

Country	Aid as % of national income, 1999 (1994)	Aid in $ per capita, 1999	Comments
Malawi	25.0 (41)	41	$446m received
Zambia	20.0 (23)	63	$623m received
Guinea-Bissau	27.0 (77)	44	Only $44m received
Mauritania	23.0 (27)	84	$219m received
Rwanda	19.0 (95)	45	Halved since refugee crisis
China	0.2 (0.6)	2	Largest receipts: $3224m
India	0.3 (0.7)	1	Second largest receipts: $2324m

The role of philanthropic foundations

Each of the organisations detailed in Table 14 is based on the initiatives of a family whose wealth was made from the key parts of the economy at the time. While these projects do have an impact, they will not solve development issues on their own. They work with established agencies, such as the UN, to avoid being seen as instruments of the developed world.

Table 14 The role of philanthropic foundations

	Rockefeller	Wellcome	William and Melinda Gates
Year of foundation	1913	1936	2000
Website	**www.rockefeller.org**	**www.wellcome.ac.uk**	**www.gatesfoundation.org**
Purpose	Use of wealth for the good of others	Research to improve human and animal welfare	'There is no greater issue on earth than the health of our children' (W. Gates, 9 May 2002)
Source of wealth	Oil	Pharmaceuticals	Software developments
Projects	• Sustainable energy • Children's Vaccine Initiative with UNICEF and WHO • Rice biotechnology • Population, female education and contraceptive policies in Africa • AIDS vaccine initiative	• Research on malaria and the Multilateral Initiative on Malaria • The Human Genome Project • Much work is through funding scientific education and research	• Developed out of Gates Learning Foundation, providing access to technology • Merged with Global Alliance for Improved Nutrition (GAIN) in 2000

Third World debt

Much debt is the result of falling commodity prices so that LEDCs are unable to repay loans. The IMF imposed **structural adjustment programmes (SAPs)** that were dependent on withdrawing state intervention in the economy and liberalising trade. These further increased debt. The **heavily indebted poor countries (HIPC)** initiative (1996) is an agreement to reduce the debt of 41 countries, but it is not supported by many donors, for example Germany and Japan. Only 14 countries have met the criteria for inclusion. The initiative only scratches the surface of debt and most countries do not provide the UN-recommended 0.7% of GNP as aid (UK, 0.27% in 1997). In 1999, debt was abolished in some of the poorest economies by G7 countries, partially in response to pressure from **Jubilee 2000**.

3 Can sustainable development be achieved?

The concept of sustainability

Sustainability dates from the first **Global Environmental Summit** (1972) in Stockholm. It was first expressed as environmental objectives:

- to maintain ecological processes and life-support systems

- to preserve genetic diversity
- to ensure the sustainable utilisation of species and ecosystems

The concept was developed by the Bruntland Report (1987) and defined as 'development which meets the needs of the present without compromising the ability of future generations to meet their own needs'.

The concept of **economic sustainability** follows from Bruntland because it examines the ability of economies to maintain themselves when resources decline or become too expensive, and when the populations dependent on the resources are growing.

The **Rio Summit** (1992) introduced agreements on:
- climate change (challenged by the USA at Kyoto)
- biological diversity
- Agenda 21
- forest principles
- desertification

The principles set out by the **Rio Declaration** are outlined below.

Environmental principles

- People are at the heart of concerns for sustainable development.
- States have the right to exploit their own environment, but they should not damage the environment of other states.
- Environmental protection should be an integral part of the development process.
- People should be informed and states should make people aware.
- There should be environmental legislation and standards, but these must not affect other countries.
- Laws should be enacted regarding liability for pollution and compensation.
- The relocation and transfer of activities and substances that are harmful to health should be prevented.
- **Environmental impact assessment (EIA)** should be undertaken for all proposed economic activities.
- States should pass on information about natural disasters and notify neighbours of the foreseen and accidental consequences of any activities that might cross boundaries.
- The environmental and natural resources of people under oppression, domination and occupation should be protected.
- There should be a precautionary approach to the environment in all states, according to their capabilities.

Economic principles

- The right to development must be fulfilled so as to meet equitably developmental and environmental needs of present and future generations.
- All states shall cooperate in eliminating poverty in order to decrease disparities in standards of living.

- The special needs of developing countries, particularly the least developed and environmentally most vulnerable, should be given priority.
- States should cooperate to restore the Earth's ecosystem. The developed states acknowledge the responsibility they bear for the demands their societies place on the global environment, and the technologies and financial resources they command.
- Unsustainable production and consumption patterns should be eliminated and appropriate demographic policies should be promoted.
- Scientific information and innovative technologies should be transferred to improve understanding.
- States should support an open economic system. Trade policies should not contain arbitrary or unjustifiable discrimination. Unilateral actions to address issues should give way to international consensus.
- National authorities should endeavour to promote the internationalisation of environmental costs, taking into account that the polluter should pay.

There were also **social principles** and **peace principles** in the declaration.

The sustainability dilemma

The issue is that MEDCs demand resources whereas LEDCs are supplying the resources that make MEDCs affluent. People consume organic matter to the equivalent of 40% of the primary production of the Earth's ecosystems, either as food or as primary products such as timber. Globally, people have converted 29% of the land area to agriculture and settlement (Table 15).

Table 15 Percentage of land area converted to agriculture and settlement, 2000

Region	%
Asia	44
Central America and Caribbean	28
Europe and Russia	35
Middle East and north Africa	12
North America	27
Oceania	9
South America	33
Africa south of the Sahara	25

Green growth and ecotourism: the case of Silky Oaks lodge

Silky Oaks is an ecotourist resort in the Daintree rainforest near Mossman, Queensland, Australia. It is owned by P&O, a TNC whose activities range from transport and logistics to tourism.

The promotional brochure for the resort uses the following description:

> Silky Oaks lodge is an idyllic world heritage retreat that pampers you with an air of casual sophistication. Set on the Mossman Gorge, the main lodge overlooks a rainforest lagoon that has no equal for natural beauty. Even though you are so close to nature, there is no sense of 'roughing it' while you get an armchair view of one of the most diverse ecosystems on this planet.
>
> The Daintree is the oldest living rainforest on earth, a time capsule of primitive plants and birds of paradise. A place where scientists and nature lovers discover new species of flora and fauna every year. It is a perfect base to explore this region. A luxury retreat with air-conditioned chalets and fine dining.
>
> For the early riser, there is also the opportunity to view the extensive birdlife found around the lodge. The area is home to many endemic species and you are free to explore the habitats of rainforest, river, open eucalypt woodland, and the nearby Cane Falls.
>
> No trip to this region is complete without a cruise to the Great Barrier Reef to snorkel and explore a kaleidoscope of coral and fish. Spend another day on one of the three Australian Wilderness Safari tours exploring the spectacular wet tropics of far north Queensland. Together with our expert guide, you'll unlock the secrets of the rainforest and come to appreciate the complex natural history of this ancient world.

What is ecologically sustainable about Silky Oaks?

- The guided tours are with qualified biologists and environmental scientists, using a variety of locations.
- The nature walks in the grounds and bird watching are activities that do not harm the environment.
- It is open to animals.
- It conserves rainfall for use.
- It tries to educate people on how to sustain the environment through a range of ecotours.
- Staff have all received some ecological training.
- The buildings are constructed using timber from sustainable sources.
- Waste is partly recycled, treated and sorted.

What is less sustainable?

- The site had to be placed in the forest at some cost to the environment.
- It is only reachable by road and is 75 minutes from Cairns.
- Importing the luxury food and wine is expensive.
- Air conditioning is expensive.
- Visits to the Barrier Reef are fuel expensive and are contributing to destruction of the reef.
- Waste has to be taken away from the lodge.

Questions & Answers

In this section of the guide there are three questions based on the topic areas outlined in the Content Guidance section. Each question is worth 25 marks. You should divide your time according to the mark allocation.

To answer at A2 you must be prepared to assemble your response into coherent prose consisting of sentences and paragraphs, with a brief introduction and conclusion. Most questions will be in two parts. You are expected to use the information, map, table or cartoon supplied with the question as a stimulus for your answer.

You might be asked to describe, but at this level the question is more likely to ask you to 'explain', 'state how and why' or 'give reasons for'.

Examiner's comments

Candidate responses are followed by examiner's comments. These are preceded by the icon e. The comments indicate how each answer would have been marked in the actual exam and give suggestions for improvement.

Most questions at A2 involve assessment using a system of 'Levels'.

Level 1

The answer lacks breadth and is vague. There is little supporting detail. Geographical terminology is lacking or basic. The candidate might not be answering the question set, but rather one that he or she had hoped for.

Level 2

There is some breadth and/or depth to the answer, but it might be unbalanced and not follow up certain aspects. The command words in the question are noted by implication and in the conclusion. The answer describes rather than explains when requested.

Level 3

The answer has both breadth of coverage and depth of understanding. It is relevant, precise and answers in a logical fashion. The candidate does exactly what the command words require.

Quality of written communication

Up to 4 marks are added to the paper total for the following qualities:

- the structure and ordering of the response into a logical answer to the question
- appropriate use of geographical terminology
- punctuation and grammar
- the quality of spelling (if you are dyslexic, seek special consideration)
- spelling and significant grammatical errors in the candidates' answers are underlined

Migration

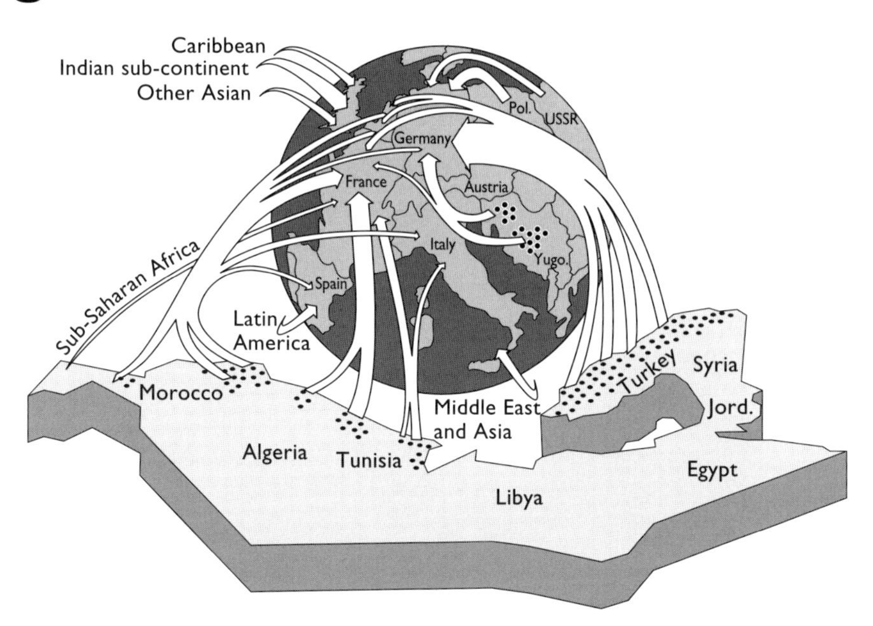

Immigration to the European Community from outside EC borders

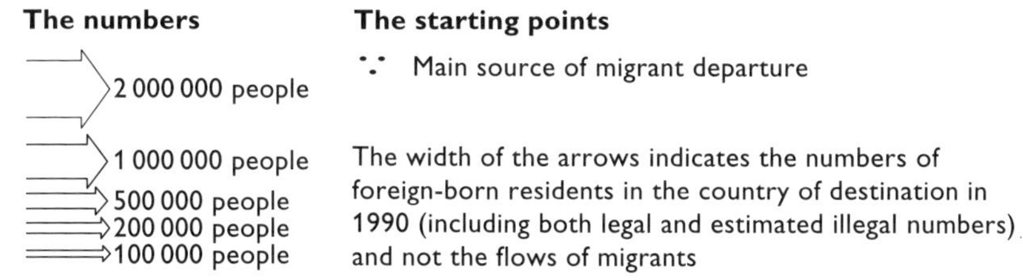

Figure 1 The origin of foreign-born residents in the EC up to 1990

(a) Suggest reasons for the pattern of migration shown in Figure 1. (9 marks)

(b) With reference to a range of examples, assess some of the benefits and costs of international migration to both source and host countries. (16 marks)

■ ■ ■

Answer to part (a): C-grade response

(a) Figure 1 shows massive migration from a group of countries, with the largest migration coming from Turkey. This was particularly evident between 1945 and 1989 when males moved to West German cities such as Stuttgart. Up to 2 million people have voluntarily migrated to Germany for a better lifestyle. The Turks were

unskilled labour and cheap to employ in Germany. The main reason for moving was to earn money and build up their depleted skills.

Further migration can be seen from Yugoslavia, with around 500 000 people emigrating. This could have been both forced and voluntary migration. The collapse of communist control in Yugoslavia saw mass migration and ethnic cleansing which ended in war. The Yugoslavs became refugees — this is people who don't return home because of fear of political or racial percecution. The other reason for people migrating was so that they could gain a regular wage and skills. Migration also occurred from many north African countries such as Algeria, Tunisia and Morocco. The majority of migrants moved from Morocco for a better quality of life in an MEDC. African civil wars have been responsible for mass migration of civilians, forcing them to become refugees. Also, the lack of jobs have seen many people migrate voluntarily. Over 1 million Algerians and Tunisians have migrated to France. These countries are French colonies, so the migrants speak the language and will find it easy to get jobs.

Many Vietnamese — 2 million — migrated because of war and became refugee between 1970 and 1975. Many people seek asylum because of political or racial beliefs. They also want a better quality of life.

e In the first paragraph, the answer drifts into part (b) and so will not be credited. In the second, it is often inaccurate and drifts into causes that explain some migration but not the extent implied. The final paragraph is not about the pattern shown. The answer would have received higher marks if it had not repeated itself, had kept to the question and been better organised. Nevertheless, the candidate has made use of the resource and is awarded 5 marks.

■ ■ ■

Answer to part (a): A-grade response

(a) The obvious migration trend is for people from less economically developed countries, LEDCs, to go to more economically developed countries — MEDCs such as France, Britain and Germany. The main reason for this is due to economic opportunities in the country. This relates to the Lee model of migration — where opportunities are greater, the migration is greater.

Many Caribbeans, Indians and Asians came to Britain because after the Second World War there was a labour deficit and, being part of the British empire, they were propagandised into working in the 'mother country', Britain. Over 150 000 have come so far as the economic opportunities are greater, with schooling, higher wages, better quality of life and a developed infrastructure. In their own countries there was poor infrastructure, poor standards of living and high unemployment, which all resulted in mass migration to Britain, which needed to rebuild its economy. This has carried on for many years as the economy has grown. The source countries have not grown and have suffered famine, political unrest and poor employment. It is the same for north African countries, such as Morocco,

Algeria and Tunisia, and Turkey; the economic opportunities are much greater in Europe and the migrants send remittances back to their families at home which will in theory help development back home. Also, the countries were historically colonies of France, which needed to recruit a workforce, so migration was possible. The cultures are quite similar, with the same language, such as Latin America to Spain.

Unrest and war also drives people out, often as refugees, as has been the case in the former Yugoslavia. However, here people left to work in Germany before the civil wars and ethnic cleansing, in order to find work.

e The candidate gives at least four reasons, supported by data and examples, for 8 marks. The answer also refers to theories. It does not get full marks because some of the ideas become generalised and there are some inaccuracies. However, the candidate uses the resource well. The quality of written communication is of a candidate who might gain 4 QWC marks overall.

■ ■ ■

Answer to part (b): C-grade response

(b) There are many benefits and costs of international migration. Mexicans moved to America during the 1960s in order to find jobs. America's southern states were in a period of rapid economic growth. Texas, California and Los Angeles all had jobs available in factories or as cleaners. These jobs were poorly paid and dirty, but they were happy to take them. Mexico had 40% unemployment and it took pressure of local resources. The Americans were able to benefit from cheap labour. However, the women had to stay in Mexico and look after the children, families and farmland. In the 1970s, a worldwide recession hit which caused major tension between the Americans and Mexicans. The Americans had a high rate of unemployment and wanted the less desirable jobs back. A ban was placed on Mexicans entering the country because of the problems that were being encountered. In 1990, 12 million Mexicans were in America, 10% of whom were illegal immigrants. There were benefits to both the host and source countries, but the policy had major problems which America found hard to solve.

Between 1945 and 1949, many males from Turkish villages migrated to West German cities. The cities of Stuttgart and Pforzheim offered cheap jobs which were dirty to the Turks. They were happy to take the jobs because they seeked a regular wage. There were benefits to both the source and host country. Germany were able to solve their labour shortage by bringing in cheap labour. Business benefited from increase profits as a result. The Turks were able to gain a regular wage, which wasn't possible back home. Also, pressure was took off the local resources in Germany. However, some major problems did result — the Turks formed a negative ethnic group which caused tension amongst the Germans. When the German unemployment rate rose, racial tension was again shown to the Turks. This was due to the unification of Germany. When people got to a leaving age, they left

Turkey for employment in Germany, and skilled people left. The Turkish migrants gained skills which would be useful if they returned to Turkey. A cheap, skilled workforce in Turkey may be able to attract investment by TNCs.

The migration of people has both benefits and costs. These were seen in Yugoslavia and Vietnam. The migrants were able to leave an area of war but at the same time they were put in overcrowded refugee camps across the world. This put pressure on the host countries. Also, many of these refugees seeked asylum in countries such as England. The benefit to the asylum seekers may be great for them but not for England. Free health care and education will go to the migrants.

e The first paragraph is descriptive — but note that jobs were often in agriculture as well. There are more points in the second paragraph, but the actual nature of the costs and benefits is not clear to the examiner. The final paragraph has no evidence of costs or benefits. The answer shows no attempt to follow the command words and therefore gains just 8 marks. The quality of written communication is of a 3-mark standard.

■ ■ ■

Answer to part (b): A-grade response

(b) Migration is good for the source country because it relieves already exhausted resources. It also improves the culture of many adult males and gives them skills that can be used when they return to the source country. Salutatan, Turkey is a major source town for migration to Germany, to towns such as Pforzheim. They gain money to send back. A recent study in Pakistan discovered that remittances help development. In Salutatan, remittances can be spent in the shops or on agricultural development and invested to help the country's economy. Malthusian theory tells us that in the supply countries, the absence of the males will help keep the population increases down. The movement of Mexicans to the USA gave others the chance of jobs in Mexico.

A benefit for a host nation like the USA is that the Mexicans add to the workforce and will do jobs, such as fruit picking in Texas and California, that the Americans do not want. Immigrants can also bring skills, such as computer programmers from India in Britain. In Britain, Indian food has become very popular. Immigrants enable people to widen their cultural horizons and appreciate different languages and religions. A multi-cultural society developes, which is an important factor in globalisation. The people often take over run-down areas and rehabilitate them. They pay taxes, which helps the national economy. In Germany, because they work they are not a drain on the social security.

The costs to the source countries is a loss of labour and skills and enterprise which could be vital for development. An emigration culture can grow up which in Salutatan stifles development. Ireland also has an emigration culture. This gets worse as temporary migrations become permanent and development is slowed. The birth rate is slowed and some towns do lack children as only the old are left. They are reliant on remittances. With fewer people, the economy of a country can be slowed so much that it does not develop. Sending males is not a sustainable solution.

The cost to the host nation is that it puts pressure on the infrastructure and resources and that is why many try to stop migration into California. Up to 10 000 Mexicans are caught trying to enter the USA illegally and catching them is a cost. People do resent the immigrants, especially if they think that the immigrants have taken their jobs. In extreme cases, this can result in racial tensions being raised, especially by right wing parties. Racial issues become worse if people congregate into ghettos, such as Brixton. There can be language problems for immigrant workers and other cultural differences that can result in misunderstandings.

e While not perfect, the answer does group the costs and benefits into paragraphs and there is support for the breadth of ideas from a range of places. There is no conclusion. It gains 14 marks. If the rest of this paper was of a similar standard, it would gain 3 QWC marks.

Uneven development and globalisation

Figure 1 The level playing field

(a) Suggest how the 'unlevel playing field' between the North and the South developed. (9 marks)

(b) With reference to *one or more* named TNCs, examine their role as global producers and employers. (16 marks)

Answer to part (a): C-grade response

(a) There are several ways in which this uneven playing field developed. These include: the initial advantage of some countries; entrepreneurship of citizens; climate; disease; and topographical factors such as relief and stability of the land.

Some countries have a natural initial advantage over others. Britain is a good example. We are not suceptable to natural disasters such as earthquakes or flood or volcanoes. We have a rich supply of natural resources to provide for our population. All of these factors have enabled us to build up a firm, stable infrastructure from which we can develop. The entrepreneurship of the original settlers in countries have provided a development platform. The people may have been risk-takers who decided that they should mine for profitable resources such as coal. If a country is to develop, it must have a stable trade platform where it can export goods and import goods it cannot naturally produce or harvest. Many countries, such as Bangladesh and Rwanda, have had little to offer in the way of exports and suffer today in the global market place. Other countries have had resources such as tea (India), coffee (Brazil) and minerals such as gold (South Africa).

A country's climate is important. In the northern hemisphere we have hospitable climates that encourage farming. Other countries do not have the environmental factors such as good climate and fertile soils to aid development. Low lying countries such as Bangladesh find difficulty setting up an infrastructure because of flooding, whereas in France the land is less likely to flood and is fertile.

As I have illustrated, the North–South gap has not developed recently. People living in the North are lucky in that their host country has had some massive initial advantages over those in the South. These advantages have been built upon and utilised to create the development gap we witness today.

e The answer contains some ideas about initial advantage but is too deterministic about the role of physical factors. The candidate under-uses the cartoon and therefore misses the issues of changing technology and the role of the WTO. This answer gains 5 or 6 marks.

■ ■ ■

Answer to part (a): A-grade response

(a) There are several reasons why the playing field has become less level, as illustrated by the cartoon.

World trade has evolved since 1600 when the first colonists began to exploit the riches of the new world, such as gold. The local population was often enslaved in the quest to exploit the natural wealth for the benefit of the colonists. By the nineteenth century, colonial powers such as the UK and France used their military power to assist in the exploitation of primary products which were returned to the mother country for processing and manufacture. These were sold back to the South as costly manufactured goods. Exploitation was aided by mechanisation and the capital to develop mines (from earlier exploitation of resources).

This pattern continues today as neo-colonialism, where it is the TNCs in the developed North which now exploit the resources in the former colonies such as Zambia and India. This pattern has been assisted by the World Trade Organisation which has written rules making it easy for companies to exploit the Third World. As a result, the South has fallen further behind and has had to borrow money to aid development and pay it back from its exports of resources. Debts make it even harder to earn money and to trade fairly.

This is a good historical view, with some support, but the points made need more evidence. It makes good use of the cartoon as a stimulus. The theme is historical and the answer recognises this. It scores 8 marks, while the QWC is of a high standard.

■ ■ ■

Answer to part (b): C/D-grade response

(b) My TNC is Microsoft and I could have chosen a number of other MNCs (multinational companies) such as GAP, Nike or Coke.

Microsoft are the world's largest computer firm. They have branches in many locations throughout the world, in MEDCs and LEDCs. LEDCs are desperate to attract large MNCs as they see it as a way of attracting other investors in the hope of further development. The LEDCs have a massive potential labour force for these companies. A large company like GAP may employ a whole town or village in the area where it is located. MNCs like employing low-skilled workers as they do not have to pay them anywhere near a decent wage. In high-tech industry, more skilled workers are needed, so setting up a Microsoft factory or office may not be suitable if adult literacy is low. When offices are set up in LEDCs, the companies are treated like royalty. They are taxed at low rates and are not pressurised about things like environmentally damaging factories. To sum up, most MNCs which set up factories or offices in LEDCs look like they are doing a good thing by the country, whereas they may be just using it and exploiting its people.

In a global market place, large company branches are a valuable commodity for a country. They are rarely pressured about the way they treat employees or the harmful effects to the environment. This means that the large companies practically control the countries they are set up in and this can be dangerous. A countries government will not challenge them for fear of the relocation of the branch firms to another country, or a massive cut in the production.

The answer names TNCs but does not include any detail about them and where they operate. It outlines branch plants and their effects, and it says why they produce in LEDCs and a little about employment, but does not offer details. There are slips of grammar and spelling, while expression is poor, which would affect the overall QWC mark. This answer is awarded 7 marks.

■ ■ ■

Answer to part (b): A-grade response

(b) TNCs are global producers and employers. Many have factories all over the world, for example CEMEX. A firm from Mexico has factories in Latin America, Spain and other countries. They import their technology from Denmark and export the goods worldwide. Therefore, they employ people worldwide.

Many electrical goods are made in Japan (Panasonic), Singapore (Siemens) and the Far East, yet people in the UK are able to buy them. This is because TNCs such as Daewoo (Korean) and Bosch (German) are well-known global producers. They export their brand everywhere.

LG, a Korean firm, has set up a factory in south Wales. This has benefited south Wales greatly as it had experienced deindustrialisation and there was a great deal of unemployment. LG has employed a large number of Welsh, but it is not all good news. The Welsh are being exploited. The Koreans get paid more than the Welsh do. However, the investment from LG has helped to encourage more investment from other TNCs, for example Ford. This investment multiplier is helping to reindustrialise south Wales. More people are employed and dependency on those in work is lowering. The quality of life improves and people spend more in the shops, so more people gain money.

However, TNCs have the money and power to pull out of any investment they have. This occurred with the American Campbell's factory in Maryport, Cumbria. The company said it pulled out to reinvest the money into other factories but it had devastating effects on the small town. It left well over 1000 people unemployed with reduced quality of life. This would have left the town with less income to spend.

Nevertheless, there is a growing number of TNCs in Britain and a growing amount of foreign direct investment. Britain is seen as a successful place to invest because companies benefit as producers from being in the EU. They have less tariff barriers to content with. Renault took a chance and invested in Romania which benefited the country, gave people employment and encouraged more investment. More people are employed when TNCs go to LEDCs, for example VW in Brazil or Nike in Vietnam. Some firms employ women, which is good.

TNCs produce goods globally and are able to export globally due to their wealth. They are also able to employ large numbers of people — Toyota employs people in many countries. However, these people may be exploited and the company may not understand cultural and social differences. All in all, most TNCs are beneficial and there is bound to be drawbacks for any company.

This answer names TNCs and includes points about their production and employment that are both positive and negative. The candidate examines the role of TNCs, as requested. The answer lacks detailed examples, but it does appreciate the consequences of TNC investment. It cites correctly the multiplier principle, but the case of Ford is wrong — it was there before LG. Nevertheless, it is awarded 14 marks, which is just about A-grade standard. It is on the road to 4 QWC marks.

Changing population structure

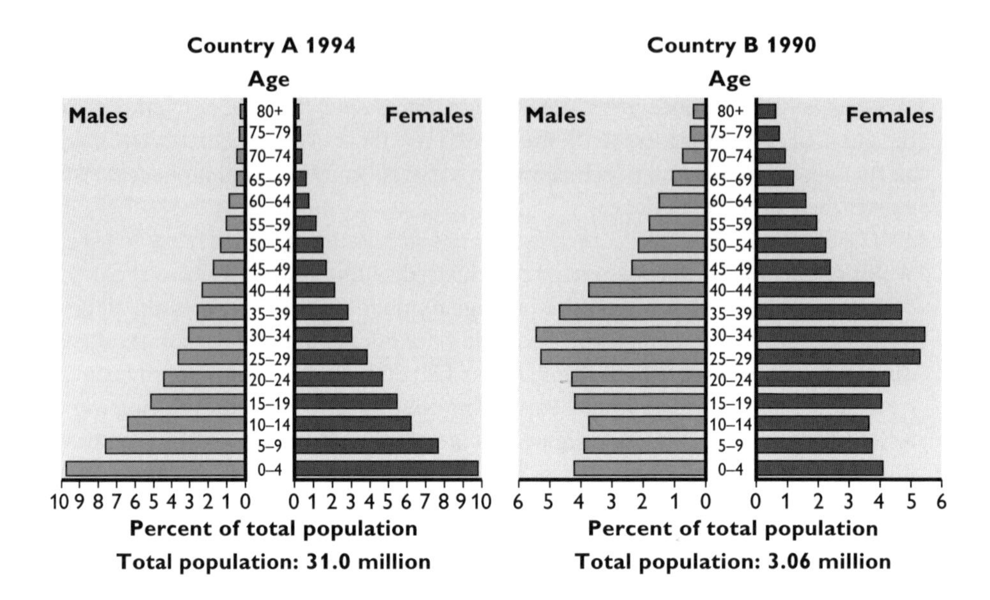

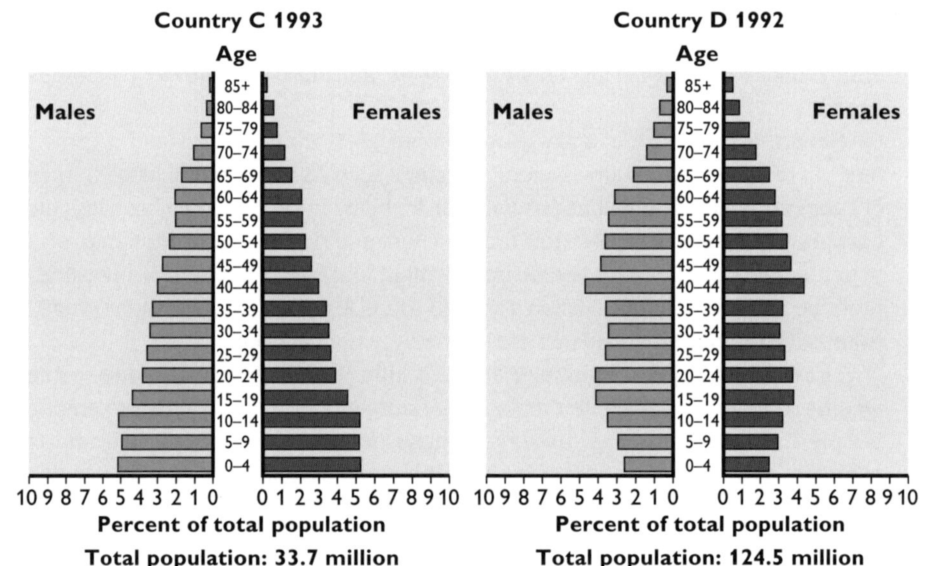

Figure 1 Age/sex pyramids for four countries

(a) Study Figure 1, which shows age/sex pyramids for four countries. Describe and suggest reasons for the variations shown in the population structure. (12 marks

(b) With reference to named examples, discuss the geographical issues arising from changing population structure. (13 marks

Answer to part (a): A-grade response

The pyramid for A shows a very youthful structure whereas that for D shows an older structure due to a greater percentage of old people in the population. Those pyramids with a broad base, such as A, have a high birth rate and a high fertility rate and less disease control, resulting in a high death rate and infant and child mortality. In comparison, B also has a high birth rate. Countries C and D have lower birth rates, possibly due to education about contraception and the changing role of women. It is probable that unlike A, there are no cultural or religious pressures to have a large family. In C and D a change in social values has occurred and lowered the birth rate; to such an extent in D that the population may not even replace itself. The empowerment of women and their emancipation (having choice of careers and choosing to follow careers rather than have children) and improvements in medical care have reduced child mortality, so that people need not have children in case some die.

In A, the higher death/mortality rate shows the high prevalence of disease, the possible lack of hygiene and sanitation as well as potentially low standards of living. In B, C and D, the improvements in medical knowledge and changed living conditions have made life less harsh. With a higher quality of life, the birth rate falls and the structure is less of a pyramid and more of an oblong.

In both B and D, the bulge in the working age groups might be due to immigration, perhaps because there is a demand for workers. B looks like an NIC, whereas D is probably more like Italy. A is probably ravaged by war, disease and famine and is an African LEDC. Pyramid C is almost perfect and could be a country that is not as developed as Italy.

> e This is a good response because the candidate attempts to account for the variations in the pyramids rather than describe each one. It could be better organised, but that is difficult in 15 minutes. It explains why there are differences and even suggests countries as examples. It will not get full marks but merits 10 or 11 given the time. 4 QWC marks would be awarded for answers of this standard.

■ ■ ■

Answer to part (a): C/D-grade response

Country A — within the elderly, there are more females, however, this is unsurprising as females live longer. This is a population pyramid typical of an LEDC, as there is a high birth rate, however, there is evidence of a high infant mortality rate, with a significant drop from 0–4 to 10–14, which could be because between ages 0 and 10 many die.

Overall, there is a low life expectancy compared with that of a richer country. This LEDC gives birth to many more children because it has traditional and religious reasons for not using contraceptives. They also see many children are a status symbol and need many children to work and bring in wages. Brides are generally younger in LEDCs, so there is more time for them to have large families.

Country B — This shows a significant baby boom, which occurred 30–34 years earlier. This is possibly a population pyramid of Britain with the baby boom occurring just after the war. Another possible reason for more females than males above the age of 34 could be because many men were killed at war.

After the baby boom there was a decrease of birth rate 30 years ago and 10 years because people were more secure in the knowledge that their sons and daughters were not going off to war and so they had less children. Another reason for this could be that since women's liberation, more women are becoming career-minded and don't feel the pressure to get married and have children. Also, the divorce rate has gone up since then, having another impact on birth rates as single people are less likely to have children. Surveys have shown that if women have the choice of how many children they have, it is considerably less than the existing birth rate.

Country D — this is more like a population column and could be somewhere like America (definitely an MEDC) with a large total population.

Yet, like country B, it shows a declining population due to a decrease in birth rates. This could be because of a rise in abortion numbers or because more are using contraceptives than 40 years ago or the fact that people are more career-minded and less family-minded. Also, people in MEDCs recognise the fact that children are expensive.

Countries B, C and D all show a larger elderly percentage than A. As they are most likely MEDCs, they have better health care and so live to an older age (high life expectancy). In MEDCs, the labour is more tertiary compared with LEDCs. LEDC workers do more hard labour, which brings down the life expectancy.

e This is a very long answer, leaving this candidate with little time for part (b) below. The country by country approach does not give sufficient attention to the reasons for the variations. Nevertheless, there are explanations in the text, even if they are repetitive and not very clearly expressed. Country C is hardly mentioned because of time. Some of the reasons are too simplistic and Euro-centric, especially those regarding LEDCs. This gains 7 marks, while 3 QWC marks would be awarded for answers of this standard.

■ ■ ■

Answer to part (b): A-grade response

A changing population structure impacts upon the social, economic and environmental climates of a particular area.

Environmentally, an increase in the population, due to both natural increase and immigration, will cause the structure to expand, as in the case of Singapore. It places pressure on the environment due to further requirements for food and water, and an increase in the outputs of waste and possibly pollution. Locations such as the Congo have high birth rates and high immigration, in this case due to the problems in Rwanda where there is civil strife and political instability. This swells the lower part of the population pyramid and puts pressure to increase food production, often increasing the overuse of fragile environments and resulting in soil degradation, soil erosion and soil exhaustion. The larger numbers exacerbate this by their increased waste and pollution of both the air (fires for cooking) and water (runoff from basic facilities).

Economically, an increase in population results in a larger workforce age group, but a lack of jobs can result in unemployment and economic downturns and a dependency of the population on the state. This happened recently in Germany where the jobs for the migrant workers who swelled the working age groups were no longer there. A fall in the birth rate and a narrowing pyramid, perhaps with outmigration, can lead to a reduced workforce, a falling output and a fall in the quality of life. This has been experienced in Russia and the southern parts of Europe. The smaller, young age groups will undermine the political strength, stability and status of these regions.

Socially, a reduction in birth rate over a long period increases the dependence of an ageing population, such as in Britain and Italy, and requires an increase in services such as meals on wheels, medical care transport and specific accommodation. In the UK, 20% of government expenditure is on such facilities. An increase in birth rate and immigration also impacts on services such as schools, childcare and transport. Washington DC will soon face a crisis due to the lack of schools and a shortage of teachers as the latest census identified a huge boom of children of high-school age, many of them Hispanics. In this case, the bulging population aged 10–16 will require language tuition.

So, changing population structures influence the social and physical geographical features of places.

e This is a sound answer, with a brief introduction and conclusion, for 12 marks. It is organised well into three groups of ideas related to the question. This is a style that would gain 4 marks for quality of written communication. The ideas are all supported by excellent case studies, which are located. The data about UK expenditure might be inaccurate but the point is made well. The flow of the answer is lost in the penultimate paragraph, which is a pity for such a high-quality answer.

■ ■ ■

Answer to part (b): E/U-grade response

The UN claims that over the next 50 years Europe's population will decrease by over a quarter. This has serious future adverse effects to the economy.

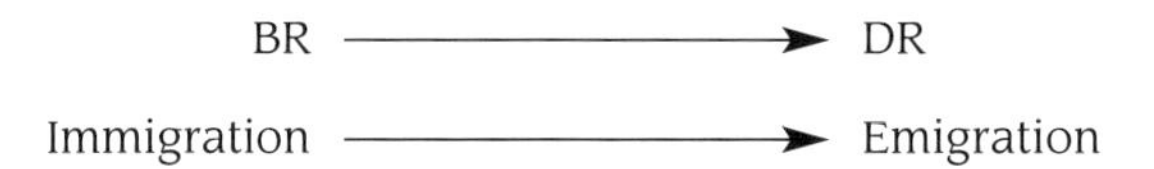

A declining population is due to less inputs and more outputs. In Europe it is because of the decrease in birth rate that the population is declining. This is due to reasons such as women working and contraception. The smaller amount of children will grow

question

up to replace the working population (15–64 years) and there will be less independent to support the dependent.

$$\text{Dependency ratio} = \frac{\text{Children + Elderly}}{\text{Working population}} \times 100$$

A decrease in dependency ratio would mean not enough people to do the jobs or to pay taxes.

An increasing population, however, also has its problems:

- lack of food
- lack of resources
- exploitation of marginal lands

This answer peters out because the candidate has run out of time. It is poorly composed and quality of written communication would achieve 2 marks at the most if all questions were answered like this one. It does have an introduction that sets the scene, but no conclusion. It does not focus very clearly on population structure, although it is implied. The supporting case studies are too general and do not help the answer. There are some issues, but they are weakly developed. This answer scores 4 or 5 marks.